Introduction to the Theory of Computation

This is a volume in
COMPUTER SCIENCE AND APPLIED MATHEMATICS
A Series of Monographs and Textbooks

Editor: Werner Rheinboldt

A complete list of the books in this series appears at the end of the volume.

Introduction to
THE THEORY OF COMPUTATION

ERWIN ENGELER
University of Minnesota

ACADEMIC PRESS   New York and London
A Subsidiary of Harcourt Brace Jovanovich, Publishers

COPYRIGHT © 1973, BY ACADEMIC PRESS, INC.
ALL RIGHTS RESERVED.
NO PART OF THIS PUBLICATION MAY BE REPRODUCED OR
TRANSMITTED IN ANY FORM OR BY ANY MEANS, ELECTRONIC
OR MECHANICAL, INCLUDING PHOTOCOPY, RECORDING, OR ANY
INFORMATION STORAGE AND RETRIEVAL SYSTEM, WITHOUT
PERMISSION IN WRITING FROM THE PUBLISHER.

ACADEMIC PRESS, INC.
111 Fifth Avenue, New York, New York 10003

United Kingdom Edition published by
ACADEMIC PRESS, INC. (LONDON) LTD.
24/28 Oval Road, London NW1

LIBRARY OF CONGRESS CATALOG CARD NUMBER: 72-88351

AMS (MOS) 1970 Subject Classifications: 68-01, 68A05, 68A25, 68A30

PRINTED IN THE UNITED STATES OF AMERICA

Contents

Preface ... vii

Chapter 1. Introduction to Finite Automata

1. Words and Word Functions ... 4
2. Symbols and Semigroups ... 9
3. Production and Acceptance ... 15
4. Grammars and Automata ... 24
5. Regular Sets ... 37
6. Building an Acceptor ... 45
7. Finite Transducers, Minimalization ... 55
8. Truth-Functions ... 69
   References ... 84

Chapter 2. Recursive Functions and Programmed Machines

1. The Universal Calculator ... 90
2. Examples of Computable Functions ... 100
3. The Structure of Programs ... 114
4. Computable Functions ... 127
5. Loop Programs and Primitive Recursive Functions ... 137
6. Complexity and Growth of Primitive Recursive Functions ... 143
7. Stored Programs ... 149
8. The Thesis of Church and Turing ... 160
9. Random-Access Stored Program Machines ... 172
   References ... 189

Chapter 3. **Elementary Syntax**

1. The Structure of Language — 195
2. Introduction to the Theory of Context-Free Grammars — 199
3. Memory Management — 217
4. Unsolvability of Grammatical Problems — 223
   References — 226

Index — 227

Preface

The goal of this book is to provide the student with a thorough understanding of the fundamental ideas that enter the study of computing processes in general.

The material developed from a junior level course at the University of Minnesota, given to undergraduates in mathematics, computer science, or electrical engineering curricula.

The book assumes virtually no formal background except for a general acquaintance with modern mathematical formalism and concept formation—obtained as a matter of course in the first two years of studying one of the above disciplines. Familiarity with programming will help but is not necessary.

We are careful to give extensive motivation for the notions that we introduce and provide many and various exercises to train the student in making the transition from concrete experience with machines and computations to the corresponding abstract notions.

The first chapter, on finite automata, is largely traditional; a special feature is the insistence on the interconnections between automata and languages. This topic is taken up again, with added motivation, in the last part of the book.

The main novelty of this book is the treatment of recursive function theory. The basic notions are seen to flow naturally out of a programmed computer concept. This contact between recursive functions and programs is stressed and maintained throughout the development of recursive function theory, and beyond. This results in a number of conceptual and technical simplifications that make the subject more easily accessible. Moreover, our approach enables us to motivate, and treat to some extent, some important topics such as complexity theory and the theory of self-modifying programs.

It is a pleasure to acknowledge my indebtedness to my colleagues E. G. Fuhrken, R. Jeroslow, and W. Richter who have taught the course from my notes and provided me with detailed criticism and suggestions. The definitive book on the theory of computing has yet to be written; for this the subject is too young. The present book grew out of an immediate teaching need and hopes to satisfy it in a modest fashion.

A word on the literature in the field is in order. For most of the basic results, there exist ample references in the textbook literature. We shall not give credit to the scientist who invented these results except in those circumstances where the theorem in question is customarily referred to as X's theorem. Material that has not found its way into most textbooks is credited to the appropriate authors and references given.

CHAPTER

Introduction to Finite Automata

In the development of a mathematical theory of computing, finite automata play a very important role. The ideas of a system of data given as words over a finite set of symbols and automatic computations thereon present themselves in their simplest and most manageable form when we take computations by finite automata. Thus we will understand the behavior of finite automata to a degree of effectiveness that we will not reach again for other more general models of computing devices.

What is a finite automaton? And what is it supposed to do?

Let us imagine a black box that is in the business of processing data. Thus data are fed into this box one piece of data after another. Let us call these pieces of data "input symbols." Because of the obvious technical restrictions, it is reasonable to assume that the set of symbols is finite: A finite technical device cannot, at any one given moment, distinguish between more than a limited amount of different input pieces; as soon as there are too many it will "confuse" them. At any one given moment, the black box itself is in some "state." By this we mean the description of the condition of all the relevant positions (such as voltage against ground, open-, or closed-ness of switches, etc.) of the parts (such as pieces of wires, transistors, switches, etc.) that make up the innards of the box. Again, it is reasonable to assume that the set of states of the black box is finite. Of course, the active voltage at one particular point, for example, may assume a continuum of values. However, from the point of view of processing data, voltages that are very close together will not be distinguishable by the subsequent behavior of the device; they will be submerged in "noise" and thus confused. Finally, the black box will have some output. What and when it will produce an output it itself decides; that is, outputs are associated to some of the states. Again, there will only be a finite set of pieces of output data that are associated to the various states. These pieces of data will be called "output symbols."

Our black box thus transforms a sequence of input symbols into a sequence of output symbols; that is, it computes a function (from the set of sequences of input symbols to the set of sequences of output symbols). To do automata theory thus means to discuss the kinds of functions that can be so computed. This is the goal of the present chapter.

Naturally, we have to start with the basic concepts surrounding the notion of "sequence of symbols" and functions on such sequences. This is done in Sections 1 and 2* (which contains a discussion of some finer points and may be deleted at first reading).

In Section 3 we give some motivation for restricting our attention (for the time being) to only one type of operative use of our finite computers, namely, as acceptors. This topic is then discussed in Sections 4 through 6. In Section 7 we return to the original problem (of functions computed by finite computers); Section 8* again contains optional material.

We use terms such as "finite computer," "black box," and so on, only in nonformal contexts, to evoke a certain picture in the reader's mind. The formal concepts "finite automaton" and "finite transducer" are introduced in the appropriate context.

The present chapter introduces the basic concepts of automaton theory in Sections 1, 3–6, and 7. The starred Sections 2 and 8 contain discussions of additional material and of finer mathematical points and may be deleted in a first reading.

1. WORDS AND WORD FUNCTIONS

Imagine a finite set A of *symbols*. What comes to mind is perhaps the numerals $A = \{0,1,2,\ldots,9\}$, the lower-case alphabet $A = \{a,b,c,\ldots,z\}$.

We shall call such a finite set quite generally an *alphabet*. Let A be any alphabet. Any finite sequence of symbols from A forms what we call a *word over A*. For convenience, we also include the sequence of length zero (which cannot actually be written out!) among the set of words over A; it is called the *empty word* and denoted by the letter "λ."

In general we use lower-case letters from the beginning of the alphabet to denote symbols, and lower-case letters from the end of the alphabet to denote words; upper-case letters from these parts of the alphabet denote sets of symbols and sets of words, respectively. The term A^* represents the set of all words (including λ) over an alphabet A.

If $u \in A^*$ and $v \in A^*$, we may form a word w in A^* by attaching the word v at the right end of the word u without space. (*Remark*: If we wish to use "space" or "blank," these actually would be additional symbols. To avoid confusion, the reader should therefore be sure that he understands the difference between the empty word and a blank space!)

Example. If $A = \{0, 1, 2, \ldots, 9\}$, $u = 2314$, and $v = 972$, then attaching v to the end of u results in the word $w = 2314972$.

The operation of attaching the word v to u is called *concatenation* (from "chaining together" in Latin); we represent it symbolically by

$$w = u \cdot v.$$

The dot between u and v indicates a binary operation. Note that the order of the two factors is important, for clearly $u \cdot v$ is a word different from $v \cdot u$. (In our example, the first would be 2314972, while the second would be 9722314.) The algebraic properties of this operation are investigated in Section 2 in an abstract manner. For now we simply collect the following obvious facts about concatenation.

1.1 Proposition. (a) For each $u \in A^*$ and $v \in A^*$, there is a unique word $w \in A^*$ such that $w = u \cdot v$.
   (b) For all $u \in A^*$, we have $u \cdot \lambda = \lambda \cdot u = u$.
   (c) For all $u, v, w \in A^*$, the associate law holds; that is,

$$u \cdot (v \cdot w) = (u \cdot v) \cdot w.$$

   (d) If $u \cdot w = v \cdot w$, then $u = v$—that is, "right cancellation."

The following proposition is intended to capture the idea that every word is obtained by starting with the empty word and concatenating it repeatedly with one symbol at a time. For this purpose, it is advantageous to make no distinction between a symbol and the word consisting of a single instance of that symbol.

1.2 Proposition. Suppose W is a set of words over A such that $\lambda \in W$ and such that, whenever $u \in W$ and $a \in A$, then $u \cdot a \in W$. Then $W = A^*$.

This proposition is indeed nothing but a paraphrase of the statement that A^* contains nothing but the empty word and those generated from it by attaching symbols from A to words already obtained. It is a useful paraphrase since it gives us a scheme whereby we can prove that some given statement S holds for all words $w \in A^*$. Namely, consider the set W of all w for which S is true. To conclude that $W = A^*$, which is what we want, all that is needed is to show two things, according to Proposition 1.2. First, $\lambda \in W$. Second, under the assumption that $w \in W$, prove that $w \cdot a \in W$ for all $a \in A$. The reader may be reminded of the principle of mathematical induction here; this is no superficiality, as we see in Section 2. Indeed, the above-mentioned scheme of proving S for all words in A^* is sometimes called a proof "by induction on the length of words," for obvious reasons.

In subsequent work, we are very often concerned with defining some function f from A^* into some arbitrary set. The concept of *length* is a case in point. Taking the hint from Proposition 1.2, we suggest that it is enough to define the length of the empty word and to show how to obtain the length of $w \cdot a$ from the length of w for any given $a \in A$. Therefore,

$$l(\lambda) = 0$$
$$l(w \cdot a) = l(w) + 1$$

describe the conditions which we reasonably impose on the function l. It is quite clear, even without proof, that the two conditions above do determine the intended function.

We assume here that the notion "f is a function with domain D and range R" is a basic mathematical notion that needs no further reduction to still more basic concepts. We shall often express this notion by the formula

$$f: D \to R.$$

Thus, the length function l would be indicated by $l: A^* \to N$ where N is the set of nonnegative integers.

More generally, let M be any set whatever, and let g be a given function of three variables, $g(x,y,z)$. The variable x runs over the set M; the variable y runs over the set A^*; and the variable z runs over the set A. Symbolically, we write this as

$$g: M \times A^* \times A \to M.$$

This notation is coherent with our previous notation $f: D \to R$ for functions $f(x)$ of one variable only. Namely, we consider $M \times A^* \times A$ as the set of all ordered triples $\langle m,w,a \rangle$ where $m \in M$, $w \in A^*$, and $a \in A$. Then, as x, y, z vary independently over M, A^*, and A, respectively, together they constitute one variable that runs exactly over the set $M \times A^* \times A$.

Let now $m \in M$, $g: M \times A^* \times A \to M$ be given.

1.3 Definition. If $f: A^* \to M$ is a function which satisfies the following equations for all $w \in A^*$ and $a \in A$, then f is said to be defined by primitive recursion from g:

$$f(\lambda) = m$$
$$f(w \cdot a) = g(f(w), w, a).$$

Now, the basic fact is that, whatever the functions g and the element m are, there is always such a function f, and only one:

1.4 Proposition. Let M be any nonempty set, let $m \in M$, and let $g: M \times A^* \times A \to M$. Then there exists one and only one function $f: A^* \to M$ such that

$$f(\lambda) = m$$
$$f(w \cdot a) = g(f(w), w, a)$$

hold for all $w \in A^*$, $a \in A$.

A formal proof of this proposition is more of foundational than of practical mathematical interest; it is discussed in Section 2. However, intuitively we are very easily convinced of the correctness of the proposition. For, suppose we are given m and g and any word $w \in A^*$. Let us say, for example, that $w = a_1 a_2 a_3$. Then how would we compute $f(w)$? By the second equation

$$f(w) = f(a_1 \cdot a_2 \cdot a_3) = g(f(a_1 \cdot a_2), a_1 \cdot a_2, a_3).$$

By the second equation, again, this transforms to

$$g(g(f(a_1), a_1, a_2), a_1 \cdot a_2, a_3)$$

and again to

$$g(g(g(f(\lambda), \lambda, a_1), a_1, a_2), a_1 \cdot a_2, a_3) = g(g(g(m, \lambda, a_1), a_1, a_2), a_1 \cdot a_2, a_3),$$

which can be evaluated by assumption.

The function f thus determined according to Proposition 1.3 is said to be *defined by primitive recursion from g*; the two determining equations are called the *recursion equations* for f. The concept of defining a function by recursion is quite important for the following, and we now give a number of illustrating examples.

1.5 Examples

(a) The concept of length, introduced above, fits into the scheme of definition by primitive recursion. We need to point out the function $g: N \times A^* \times A \to N$ where N is the set of natural numbers. Observe that $g(n, w, a) = n + 1$ for all n, w, and a; thus g does not actually depend on the second and third variables in this case.

(b) The function $R: A^* \to A^*$ which reverses the order of a string can be defined by primitive recursion. This function acts on $w = a_1 a_2 a_3 a_4$, for example, to yield $a_4 a_3 a_2 a_1$. The recursion equations are obviously

$$R(\lambda) = \lambda$$
$$R(w \cdot a) = a \cdot R(w).$$

Thus $g: A^* \times A^* \times A \to A^*$ is defined by $g(u, v, a) = a \cdot u$; it does not

2. SYMBOLS AND SEMIGROUPS

depend on the second variable. We adopt the convention of writing w^R for $R(w)$.

(c) Consider an alphabet A consisting of one symbol only, say $A = \{1\}$. Thus A^* consists of strings of 0, 1, 2, 3, ... strokes, and each $w \in A^*$ is fully determined by its length $l(w)$. Therefore A^* may be considered just a copy of N, the set of natural numbers; concatenation corresponds to the sum (of length) of course, not to the product as the dot might indicate. Now, for natural numbers we are quite familiar with definitions of functions by primitive recursion; in any case this is taken up in later chapters. For an example, consider the function

$$f(\lambda) = 111$$
$$f(w \cdot 1) = f(w) \cdot 1,$$

which adds three strokes to its argument. If we consider n strokes simply as a *notation* for the number n, then $f(n) = n + 3$ expresses this function. A more complex example is

$$f(\lambda) = \lambda$$
$$f(w1) = f(w) \cdot 111,$$

which computes, in the notation above, $f(n) = n \cdot 3$.

2.* SYMBOLS AND SEMIGROUPS

In the present section we treat the notions introduced in Section 1 from a more advanced standpoint. Let us consider an alphabet A and the set A^* of words over A.

We recall a few obvious facts about the concatenation operation $\cdot$ on A^*.

(a) For each $u, v \in A^*$, there is a unique $w \in A^*$ such that $u \cdot v = w$.
(b) There is a word $\lambda \in A^*$ such that $u \cdot \lambda = \lambda \cdot u = u$ for all $u \in A^*$.
(c) For all $u, v, w \in A^*$, the associative law holds:

$$(u \cdot v) \cdot w = u \cdot (v \cdot w).$$

The concatenation operation on A^* which is everywhere defined and satisfies the associative law (c), imposes on A^* a mathematical structure. Such structures are called *semigroups*.

2.1 Definition. A semigroup is a set S together with a binary operation $\cdot$ on S which satisfies the associative law. An element $1 \in S$ is called a unit element of a semigroup if $1 \cdot w = w \cdot 1 = w$ for all $w \in S$. A semigroup with unit is called a *monoid*.

Thus A^*, with concatenation as the operation and λ as a unit element, is an example of a semigroup with unit. Of course there are many other examples of semigroups that arise in mathematics, and this fact makes it profitable to study the concept in full generality. If we do this, we simply abstract from the particular circumstances that give rise to a semigroup but be content to note that the structure at hand is indeed a semigroup. This means that we do not inquire deeper into the essence of the elements and the operations, but only verify the fundamental formal properties of the composition operation. With this method, modern algebra circumvents the conceptual, foundational difficulty that arises from the question of the nature of mathematical entities. Rudely put, a modern algebraist simply says "Don't come to me with epistemological niceties about the essence of mathematical objects; I know a semigroup when I see one."

Now, do we know a "set of words with concatenation" when we see one? Clearly, a set S with composition $\cdot$ would have to be a semigroup to qualify for a set of words. But this is not enough; I'm sure that very few people would call the set of continuous functions with the operation of addition a set of words over an alphabet.

The conclusion is that we need to have a few more properties to characterize the concept. What we need to capture somehow is the idea that every word is obtained by concatenation of a finite sequence of symbols. For this purpose, it is advantageous to make no distinction between a symbol and the word consisting of a single instance of that symbol.

(d) There is a finite set $A \subseteq A^*$ such that the following is true for every set B of words over A:

Assumptions: $\lambda \in B$, $w \in B$ implies $w \cdot x \in B$ for all $x \in A$.
Conclusion: $B = A^*$.

2. SYMBOLS AND SEMIGROUPS

These additional observations on sets of words with concatenation give rise, by abstraction, to the following concept.

2.2 Definition. (a) A semigroup S with identity 1 and binary operation $\cdot$ is called a *free monoid* if there exists a set A (called the set of generators of S) which satisfies the following three axioms.
  (i) $A \subseteq S$, $1 \notin A$, $u \cdot x \neq 1$ for all $x \in A$, $u \in S$.
  (ii) If $u, v \in S$ and $x, y \in A$ are such that $u \cdot x = v \cdot y$, then $u = v$ and $x = y$.
  (iii) For all sets B, if $1 \in B \subseteq S$ and $w \in B$ implies $w \cdot x \in B$ for all $x \in A$, then $B = S$.

(b) If A is finite, then A is called an *alphabet*, its elements are called *symbols*, and S is called *the set of words over A*. The identity of S is called the *empty word*. (*Note*: If $u \cdot e = e \cdot u = u$ for all u in S, then $e = 1$.)

This definition gives us an axiomatic basis to work from; hereafter, we make exclusive use of properties expressed in, or following from, the axioms. For convenience we agree on the convention to present a monoid always in the form of a triplet of entities such as

$$\langle S, \cdot, 1 \rangle,$$

the first term being the set of elements of the monoid, the second the binary operation, and the third the unit element.

2.3 Examples and Problems
  (a) Let N be the set of nonnegative integers, $+$ the addition operation, and 0 zero. Then $\langle N, +, 0 \rangle$ is a free monoid. Prove this! What is the set of generators, and what is the unit element of this monoid?
  (b) Let F be the set of continuous real-valued functions on the unit interval and let $\cdot$ be multiplication of functions [i.e., $f \cdot g$ is defined pointwise as $f \cdot g(x) = f(x) \cdot g(x)$]. Let e be the function identically equal to 1. Then $\langle F, \cdot, e \rangle$ is a monoid. Is it a free monoid? Justify your answer.
  (c) Is there a monoid which satisfies axioms (i) and (iii) but fails to satisfy axiom (ii) of Definition 2.2?
  (d) If A_1 and A_2 are sets of generators for a free monoid $\langle S, \cdot, 1 \rangle$, then $A_1 = A_2$. This fact allows us to speak of *the* set of generators of a

free monoid; prove it! (*Hint*: Let A_0 be the set of elements $w \neq 1$ of S for which, from $x \cdot y = w$, it follows that $x = 1$ or $y = 1$. Then show that $A_0 = A_1$.)

Let us now recast the notion of recursive definition of functions in the framework of free monoids. The purpose is to define some function f from a free monoid S into some arbitrary nonempty set M. Suppose such a function is defined on the unit. Suppose furthermore that we have a way to define, once the value of f on an element w is known, the value of f on all elements $w \cdot x$ (x a generator). We are tempted to conclude that f is thereby defined on all elements. While the conclusion is true (see Theorem 2.4), the most tempting way to prove it is fallacious: Let T be the set of elements of the free semigroup $\langle S, \cdot, 1 \rangle$ on which f is defined by the stipulations above. Clearly T contains 1 and has the property that if $w \in B$, then $w \cdot x \in T$ for each generator x. Hence $T = S$. Q.E.D. This proof suffers from the same fatal error as does the following proof that all numbers are interesting: 0 is an interesting number; if n is an interesting number, then so is $n + 1$; for were it not, it would be the first non-interesting number which is a very interesting number indeed. So all numbers are interesting, aren't they? The trouble with this "proof" is that one of the basic concepts in it, the notion of interesting, is changed in the middle of the proof. The same happens in the alleged proof under discussion. The basic concept that is not rigidly imposed throughout the argument is that of a function being defined on an argument.

2.4 Theorem. Let $\langle S, \cdot, 1 \rangle$ be a free monoid, and let A be the set of generators of S. Suppose that m_0 is an element of M and that g is a function from $M \times S \times A$ to M. Then there is a unique function f from S to M which satisfies the following two equations

$$f(1) = m_0$$
$$f(w \cdot x) = g(f(w), w, x)$$

for all $x \in A$ and $w \in S$.

Proof. Of the two things to be proved, existence and uniqueness, the second is the easier. For suppose that f_1 and f_2 satisfy the equations above.

2. SYMBOLS AND SEMIGROUPS

To show equality of f_1 and f_2 we need to show, by definition, that $f_1(x) = f_2(x)$ for all x in S. Therefore, let $T \subseteq S$ be the set of all x on which f_1 and f_2 agree. Clearly $1 \in T$ by definition. Suppose that $w \in T$ and that x is a generator. Then

$$f_1(w \cdot x) = g(f_1(w), w, x) = g(f_2(w), w, x) = f_2(w \cdot x)$$

because $f_1(w) = f_2(w)$ and both f_1 and f_2 satisfy the second of the equations above. Thus $f_1(w \cdot x) = f_2(w \cdot x)$. It follows that $T = S$ and therefore $f_1 = f_2$.

To prove the existence of a function $f: S \to M$ which satisfies the equation, we have first to establish a standard for the acceptance of an infinitary entity such as a function as existing.

One possible way would be to admit only functions that are given to us by some sort of algorithm (which again would then be a concept that needs explaining) and to show, given an algorithm for g, how to construct one for f. There would remain the task of showing that the algorithm terminates with a value for each argument w from S and that the function so computed indeed satisfies the equations. None of these tasks is very simple; also, this approach really makes sense only if S has a finite set of generators. We get to this approach only much later in the course.

The second approach is set-theoretical. We consider a function $f: S \to M$ given if its *graph* $\gamma(f)$ is given. By a graph of f, we mean a subset $\gamma(f) \subseteq S \times M$ such that $\langle x, y \rangle \in \gamma(f)$ if and only if $f(x) = y$. The idea is to obtain $\gamma(f)$ as the union of an infinite increasing set of subsets of $S \times M$, each of which is a graph of a function defined on only part of S. Since the details of the proof would not contribute to our understanding, we omit them and pass on to some concepts that can easily be explained here and *do* reappear in our later work.

Let $\langle S, \cdot, 1 \rangle$ and $\langle T, +, 0 \rangle$ be two monoids (not necessarily free). A function

$$h : S \to T$$

is called a *homomorphism* if $h(1) = 0$ and $h(x \cdot y) = h(x) + h(y)$ for all $x, y \in S$. Furthermore, if h is one-to-one and onto, then h is called an *isomorphism*. [The function h is called *one-to-one* if from $h(x) = h(y)$ it always follows that $x = y$; it is called *onto* if for each z in T there is x in S with $h(x) = z$.] Two monoids between which there exists an isomorphism are called *isomorphic*.

2.5 Theorem. Let $\langle S, \cdot, 1 \rangle$, $\langle T, +, 0 \rangle$ be two free monoids; let A be the set of generators of S, and B the set of generators of T. Suppose that there exists a function which maps A onto B in a one-to-one manner. Then the two monoids are isomorphic.

Proof. Let $f_0 \colon A \to B$ be one-to-one and onto. By Theorem 2.4, there exists a function f such that

$$f(1) = 0$$
$$f(w \cdot x) = f(w) + f_0(x)$$

for all $x \in A$ and $w \in S$. Since f_0 is one-to-one and onto, it has an inverse; there is a function $g_0 \colon B \to A$ such that $g_0(f_0(w)) = w$ for all $w \in A$. Let g be the function from T to S which satisfies

$$g(0) = 1$$
$$g(w + x) = g(w) \cdot g_0(x)$$

for all $x \in B$, $w \in T$. Both f and g are one-to-one and onto. This follows from the fact that $g(f(w)) = w$ for all $w \in S$. This latter statement is proved easily by considering the set $U \subseteq S$ of those $w \in S$ for which $gf(w) = w$. Note that $gf(1) = g(0) = 1$, hence $1 \in U$. If $w \in U$ and $x \in A$, then

$$gf(w \cdot x) = g(f(w) + f_0(x)) = g(f(w)) \cdot gf_0(x) = w \cdot g_0 f_0(x) = w \cdot x,$$

hence $w \cdot x \in U$. It follows that $U = S$. It remains to prove that f is a homomorphism.

To prove $f(u \cdot v) = f(u) + f(v)$ for all $u, v \in S$, let us keep u fixed and consider the set H of all $v \in S$ for which $f(u \cdot v) = f(u) + f(v)$. Observe $1 \in H$; namely, $f(u \cdot 1) = f(u) = f(u) + 0 = f(u) + f(1)$. Suppose $v \in H$ and $x \in A$. Then

$$f(u \cdot (v \cdot x)) = f((u \cdot v) \cdot x) = f(u \cdot v) + f(x) = (f(u) + f(v)) + f(x)$$
$$= f(u) + (f(v) + f(x)) = f(u) + f(v \cdot x),$$

which shows $v \cdot x \in H$. Thus $H = S$, which was to be proved.

2.6 Remarks. (a) We have already proved [in (d) of Problem 2.3] that a free monoid has a well-determined set of generators. Theorem 2.5 now allows us to speak of *the* free monoid generated by a set of generators. If A is a set (of generators), we shall denote by A^* the free monoid generated by A.

(b) The property of free monoids formulated in Theorem 2.5 can be used to characterize free monoids; in fact, this is the usual characterization of free algebraic systems in modern algebra. [See problem (e) below.]

2.7 Problems (continued)

(e) A monoid $\langle S, \cdot, 1 \rangle$ with identity is called free if there exists a set $A \subseteq S$ such that, for every monoid $\langle T, +, 0 \rangle$ and every function $h_0 \colon A \to T$ there exists a unique homomorphism $h \colon S \to T$ such that $h(x) = h_0(x)$ for all x in A. Prove that the old and the new notions of free monoid coincide.

3. PRODUCTION AND ACCEPTANCE

Let us now return to our intuitive description of the structure and operation of finite computers in the introduction to this chapter. Imagine a black-box computer being fed sequences of digits, forming a number on which the computer then operates; or of two such sequences of digits, separated by a comma, say, which the computer then proceeds to process, for example, by performing an addition. Or even further, the sequence may be a sequence of letters and digits forming an instruction or program which the computer then proceeds to execute. In short, among all possible sequences of input symbols to the computer are those that are *meaningful* to it. The totality of all meaningful inputs thus forms a subset of all words over the input alphabet, something akin to a "language" which the computer is able to "understand." Our first goal now is to get our hands on a suitable concept of language—suitable in the sense that it is meaningful, that is, acceptable to the finite computers we have in mind.

Let us start by looking at the language that is best known to us, English. To simplify matters let us take as "alphabet" the set of all words of the English language (including their declensions, conjugations, etc.). A "word" over this alphabet would then be an utterance such as "A man is a man," but "words" would also include gibberish such as "Nice not never new." In some fashion, the meaningful phrases are distinguished from the meaningless. We discuss here two mathematical models which indicate in what way such a distinction could be understood. One is that meaningful phrases are *produced* according to some inherent laws of *grammar*. The other posits that the acceptance of an utterance may perhaps be likened to a mechanical process or machine which accepts meaningful and rejects meaningless utterances.

Among the uses to which we put colloquial English is to serve communicate findings or opinions in the form of statements of facts ("descriptive sentences"). Another is to convey instructions on how to perform certain sequences of operations such as the assembly of a tricycle or the computation of the roots of a quadratic equation ("imperative sentences").

Statements and instructions are meant to be understood. Our understanding depends on our ability to analyze such utterances, that is, to recognize them as being composites of basic (and immediately understandable) statements or instructions. The exact manner in which correctly formulated utterances are composed of their basic constituents is a matter of grammatical analysis, a topic to which we now turn our attention.

Suppose we were to make up an English sentence. We could conceivably proceed as follows. We start by putting the symbol σ on the blackboard. We shall erase it presently, but we have put it there to declare our intention to produce a sentence. A sentence is generally composed of a noun phrase and a verb phrase, in that order. Let us choose symbols $\boxed{NP}$ and $\boxed{VP}$ for these grammatical entities; erase σ and replace it by the word $\boxed{NP | VP}$.

Rewrite rule 1:
$$\sigma \to \boxed{NP | VP}.$$

3. PRODUCTION AND ACCEPTANCE

A typical noun phrase is an article followed by a noun. Thus we would again erase $\boxed{NP}$ and replace it by the word $\boxed{T|N}$, the symbol $\boxed{T}$ standing for articles, $\boxed{N}$ for nouns.

Rewrite rule 2:

$$\boxed{NP} \rightarrow \boxed{T|N}.$$

A typical verb phrase consists of a verb, symbol $\boxed{V}$, followed by another noun phrase. This gives

Rewrite rule 3:

$$\boxed{VP} \rightarrow \boxed{V|NP}.$$

Using these three rewrite rules, we produce in succession

$$\sigma \Rightarrow \boxed{NP|VP} \Rightarrow \boxed{T|N|VP} \Rightarrow \boxed{T|N|V|NP},$$

and finally we have on the board the word

$$\boxed{T|N|V|T|N}.$$

What remains is to replace the symbols $\boxed{T}$, $\boxed{N}$,..., by articles, nouns, and so on. For example

$$\boxed{the | cat | ate | a | mouse}$$

makes use of the following rewrite rules:

$$\boxed{T} \rightarrow \boxed{the}$$

$$\boxed{T} \rightarrow \boxed{a}$$

$$\boxed{N} \rightarrow \boxed{cat}$$

$$\boxed{N} \rightarrow \boxed{mouse}$$

$$\boxed{V} \rightarrow \boxed{ate}.$$

The symbols for grammatical entities such as $\boxed{T}$, $\boxed{NP}$, and so on, have as a rule no place in a finished English sentence; they are called *nonterminal* symbols. On the other hand, the English words like ⟦mouse⟧, ⟦the⟧, and so on, are here also considered as single symbols; we call them terminal symbols.

We do not embark here on the probably futile quest for a full system of grammatical rules for the English language. Rather, we restrict ourselves to treating artificial languages such as they arise in various parts of mathematics, in particular in logic and programming. (See in particular Chapter 3.) For the present, we are content with a rather small fragment of such languages, to wit, notation systems for natural numbers. The more general concept of a *grammatical production system*, which captures the structure of our English-language example above more completely, is explained in more detail in Chapter 3.

Consider an alphabet consisting of the ten numeral symbols of our usual decimal notation, $T = \{0,1,2,\ldots,9\}$. Not all words $w \in T$ are notations for natural numbers as usually understood, for example, 006, 000, 0100, and so on. To single out the set of proper notation in the spirit of productions, let us introduce two nonterminal symbols, S and A, and the following rewrite rules:

$$S \to 1A \quad A \to 0A \quad A \to 0 \quad A \to \lambda$$
$$S \to 2A \quad A \to 1A \quad A \to 1$$
$$\vdots \qquad \vdots \qquad \vdots$$
$$S \to 9A \quad A \to 9A \quad A \to 9$$

We have, therefore, two alphabets present, the *terminal alphabet* $T = \{0,1,\ldots,9\}$ and the *nonterminal alphabet* $N = \{S,A\}$; in the latter, there is a distinguished symbol S called the *start symbol* with which all productions have to start. As an example, here is the fashion in which the numeral 1492 is produced:

$$S \Rightarrow 1A \Rightarrow 14A \Rightarrow 149A \Rightarrow 1492.$$

Observe that we use the double arrow $\Rightarrow$ to distinguish between a step in the production, say

$$14A \Rightarrow 149A,$$

3. PRODUCTION AND ACCEPTANCE

and the rewrite rule applied for this step, to wit

$$A \to 9A.$$

Observe that the last 11 rules,

$$A \to 0, \quad A \to 1, \quad \ldots, \quad A \to 9, \quad A \to \lambda$$

could be replaced by just one rule, namely

$$A \to \lambda$$

(where λ denotes the empty word) *without* changing the set of numerals that can be produced starting with S.

We see at once that the set produced by the rules above is the set of all numerals in the usual decimal notation. Let us consider a second example.

Suppose that we were to produce a number which is divisible by 3. In the spirit of the production systems above we would proceed as follows. The result of a production is to be a natural number in decimal notation, thus the terminal alphabet consists of the numerals 0 through 9. (To keep the example simple, we readmit as proper numerals expressions such as 001, 00, etc.) Corresponding to the grammatical entities such as verbs, and so on, we have here the notion of a remainder after division by 3; R_0, R_1, and R_2 are the nonterminal symbols for the three respective classes of numbers. If we are to produce numbers divisible by 3, then R_0 is our choice for a start symbol, and the following will be our production rules.

$$P: \begin{cases} R_0 \to 0, 3, 6, 9 \quad \text{(which abbreviates} \quad R_0 \to 0, \; R_0 \to 3, \\ \qquad\qquad\qquad\qquad\qquad\qquad\qquad\quad R_0 \to 6, \; R_0 \to 9); \\ R_1 \to 1, 4, 7; \\ R_2 \to 2, 5, 8; \\ R_0 \to 3R_0, 6R_0, 9R_0, 2R_1, 5R_1, 8R_1, 1R_2, 4R_2, 7R_2, 0R_0; \\ R_1 \to 1R_0, 4R_0, 7R_0, 3R_1, 6R_1, 9R_1, 2R_2, 5R_2, 8R_2, 0R_1; \\ R_2 \to 2R_0, 5R_0, 8R_0, 1R_1, 4R_1, 7R_1, 3R_2, 6R_2, 9R_2, 0R_2. \end{cases}$$

The reader can convince himself easily of the fact that the set W produced by these rules is the set of natural numbers x which are divisible by 3.

1. INTRODUCTION TO FINITE AUTOMATA

The set W is what we would call a language in this context; it consists of those decimal numbers that are "meaningful" in the sense of being divisible by 3. In accordance with our general goal we ask the question as to whether there may exist a finite computer for which W is the set of meaningful, that is, acceptable, sequences of symbols.

Let us consider the example at hand. Figure 1.1 presents a method whereby we can accept or reject a number according to whether it is divisible by 3.

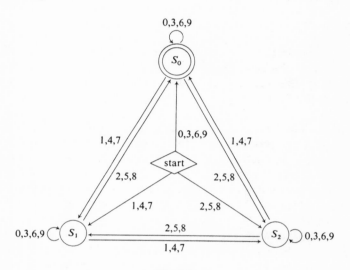

Fig. 1.1

Given a natural number, say 523281, we read this number from left to right and follow the arrows in Fig. 1.1 according to the numerals we encounter. We start in <start> and terminate in one of the vertices S_0, S_1, or S_2 of Fig. 1.1. In our example, we arrive successively at <start> S_2, S_1, S_1, S_0, S_2, S_0. We accept a number if our path terminates in vertex S_0 (this is why we circled it twice), otherwise we reject it. The proof that we accept exactly the numbers divisible by 3 is left to the reader.

Figure 1.1 illustrates a convenient way in which one can present a finite computer as it performs the task of an acceptor–rejector. This is

3. PRODUCTION AND ACCEPTANCE

what we eventually call a finite automaton. But sometimes we would prefer to visualize a finite automaton as a little machine, a black box as it were. All that can be seen from the outside is a slot and an indicator light. We know that the innards are capable of assuming finitely many different states; in some of these states the indicator light goes on, in others off. If we feed a finite sequence of symbols on a piece of tape into the slot, the following happens. At the beginning the state of the automaton is a particular state, the "start" state, which is the same for all inputs. Now the tape is read in, symbol by symbol, as it enters the box. With each incoming symbol the state of the automaton may change. This change is uniquely determined by the present state and the incoming symbol. Finally, when all the tape is read in, the automaton halts, in a well-determined state depending on the tape. If the indicator light is on at this time, then we say that the automaton has accepted the tape.

From this description it is clear that Fig. 1.1 does represent the workings of a finite automaton in the sense of a little machine. (It has four states, in one of which the indicator light is on.) Before we give the general mathematical definition of finite automata, let us consider another example.

One of the basic skills that we learn in school is to establish equations such as

$$312 + 425 = 737.$$

Can we find a production system that produces exactly the true equations of the form above? If we can, then the start symbol of the grammar would most naturally be called an *axiom* and the productions would be called *rules of proof*.

In the opposite direction, suppose that we are given equations of the form above; can we build a little *machine* that tells us whether a given equation is correct or not? We show that we can even find a *finite automaton* which is up to this job. It acts as a theorem-checking machine.

This theorem-proving, theorem-checking view of production systems and machines points to a worthwhile connection between symbol manipulation systems and mathematical logic. We return to this connection repeatedly and make use of it.

Before we present the production system, let us talk about the little machine. We cannot hope to build a finite automaton that works on the strings of symbols from $\{0,1,2,\ldots,9,+,=\}$ which are of the form in which we customarily write our equations, that is, strings like $152 + 32 = 876$. Why this is so is answered later on when we have an exact definition. Instead, we write such equations in the form

$$\begin{array}{c} 3\ 2\ 4\ 3 \\ 5\ 1\ 7\ 2 \\ \hline 8\ 4\ 1\ 5 \end{array}$$

or, more precisely,

$$\begin{pmatrix} 3 \\ 5 \\ 8 \end{pmatrix} \begin{pmatrix} 2 \\ 1 \\ 4 \end{pmatrix} \begin{pmatrix} 4 \\ 7 \\ 1 \end{pmatrix} \begin{pmatrix} 3 \\ 2 \\ 5 \end{pmatrix},$$

where we consider the columns as single symbols. Our terminal alphabet, therefore, has $10^3 = 1000$ symbols

A. THEOREM PROVING

We need only two nonterminal symbols, C_0 and C_1, representing a "carry" of 0 and 1, respectively. Here C_0 serves as the start symbol. For productions, we choose

$$C_0 \to C_0 \begin{pmatrix} 0 \\ 0 \\ 0 \end{pmatrix}, \quad C_0 \begin{pmatrix} 2 \\ 3 \\ 5 \end{pmatrix}, \ldots$$

$$C_0 \to C_1 \begin{pmatrix} 8 \\ 3 \\ 1 \end{pmatrix}, \quad C_1 \begin{pmatrix} 6 \\ 6 \\ 2 \end{pmatrix}, \ldots$$

$$C_0 \to \lambda$$

$$C_1 \to \begin{pmatrix} 0 \\ 0 \\ 0 \end{pmatrix}, \quad \begin{pmatrix} 2 \\ 3 \\ 6 \end{pmatrix}, \ldots$$

$$C_1 \to C_1 \begin{pmatrix} 8 \\ 5 \\ 4 \end{pmatrix}, \quad C_1 \begin{pmatrix} 7 \\ 7 \\ 5 \end{pmatrix}, \ldots$$

3. PRODUCTION AND ACCEPTANCE

$$C_1 \to C_0 \begin{pmatrix} 5 \\ 2 \\ 8 \end{pmatrix}, \quad C_0 \begin{pmatrix} 3 \\ 3 \\ 7 \end{pmatrix}, \ldots$$

Let us consider an example of a "proof" of an "equation":

$$C_0 \Rightarrow C_1 \begin{pmatrix} 6 \\ 6 \\ 2 \end{pmatrix} \Rightarrow C_1 \begin{pmatrix} 8 \\ 5 \\ 4 \end{pmatrix} \begin{pmatrix} 6 \\ 6 \\ 2 \end{pmatrix} \Rightarrow C_0 \begin{pmatrix} 5 \\ 2 \\ 8 \end{pmatrix} \begin{pmatrix} 8 \\ 5 \\ 4 \end{pmatrix} \begin{pmatrix} 6 \\ 6 \\ 2 \end{pmatrix} \Rightarrow \begin{pmatrix} 2 \\ 3 \\ 5 \end{pmatrix} \begin{pmatrix} 5 \\ 2 \\ 8 \end{pmatrix} \begin{pmatrix} 8 \\ 5 \\ 4 \end{pmatrix} \begin{pmatrix} 6 \\ 6 \\ 2 \end{pmatrix}.$$

This establishes 2586 + 3256 = 5842, a true equation. To show that our proposed production system does the job, we have to prove two things.

Soundness. If w is produced by the system, then w represents a true equation.

Completeness. If w represents a true equation, then w is produced by the system.

The reader can easily convince himself of these facts.

B. THEOREM CHECKING

To check a sum

$$\begin{array}{r} 2586 \\ \underline{3256} \\ 5842 \end{array}$$

for correctness, we customarily start from the right: 6 plus 6 equals 2 carry 1; we keep the carry in our memory and go to the next column: 8 plus 5, with a carry of 1, equals 4 carry 1; and so on. The carry, 0 or 1, is the only thing that we have to keep in our minds while progressing from right to left; this "keeping in mind" is realized in the automaton by a state. Of course, if once we detect an error in one column, the whole sum is wrong. Based on these remarks, it is easy to put the machine together. We have three states C_0, C_1 (for a carry of 0 or 1) and E (for error). The graph of the machine is shown in Fig. 1.2.

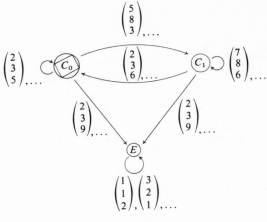

Fig. 1.2

Of the 1000 arrows emanating from each state, we have labeled only a few typical ones. Here C_0 is at the same time the start state and the only distinguished state.

4. GRAMMARS AND AUTOMATA

After the exploratory study of the previous section we are now ready to start with the proper development of our topic. The first thing to do is to give *formal* definitions for the *intuitive* notions of production systems and finite automata encountered in the previous section. Only if we have fixed notions can we hope to make useful general statements about these things. Also, when we have exactly delineated what a finite automaton is, and only then, can we *prove* that certain sets of words are *not* acceptable by any finite automaton. (By contrast, to show that some given set can be so accepted, we do not need a general notion of finite automaton; we simply construct one and see that it would be reasonable to call it a finite automaton.)

The basic question here is to characterize those sets of words that are accepted by a finite automaton. We have in mind a characterization of these sets as "languages," that is, as sets of words produced according to some grammatical rules. Only certain types of languages are so acceptable. One way to single out types of languages is by paying attention to the form of the rules in the production system. Our experience in the previous section suggests looking at the type of production rules that are listed in the definition below.[1] Instead of the term "production system" we use the more colorful word "grammar."

4.1 Definition. (a) A *right-linear* grammar $G = \langle T, N, S, P \rangle$ consists of a *terminal alphabet* T, a finite set N of *nonterminal* symbols, disjoint from T and containing the start symbol S, and a finite set P of *productions* (or "rewrite rules"). Each production is of one of the following forms

$$A \to cB, \quad \text{or} \quad A \to c, \quad \text{or} \quad A \to \lambda,$$

where A and B are nonterminal symbols (not necessarily distinct), c is a terminal symbol, and λ denotes the empty word.

(b) If x and y are words over $T \cup N$, we write

$$x \underset{P}{\Rightarrow} y$$

(or simply $x \Rightarrow y$ if P is understood), if there is a word $w \in T^*$ and a production $A \to u$ [for some word $u \in (N \cup T)^*$] such that $x = w \cdot A$ and $y = w \cdot u$.

(c) We write $x \underset{P}{\overset{*}{\Rightarrow}} y$ (or $x \overset{*}{\Rightarrow} y$ for short) if $x = y$ or there is a finite sequence $x = x_0, x_1, x_2, \ldots, x_n = y$ such that

$$x_{i-1} \underset{P}{\Rightarrow} x_i$$

for $i = 1, 2, \ldots, n$. This is often indicated by $x_0 \Rightarrow x_1 \Rightarrow \cdots \Rightarrow x_n$.

(d) The set of words w over the terminal alphabet T for which

$$S \underset{P}{\overset{*}{\Rightarrow}} w$$

---

[1] The first to connect up sets accepted by automata to sets produced by a type of grammar were Chomsky and Miller (12).

is called the set *generated* (or "produced") by the grammar G and is denoted by $L(G)$. Thus $L(G) \subseteq T^*$. We sometimes say that this set is a *right-linear language* (by abuse of the concept of language).

4.2 Definition. *Left-linear grammars* and *left-linear languages* are defined the same way as right-linear ones, with the change that productions have to be in the form

$$A \to Bc \quad \text{or} \quad A \to c \quad \text{or} \quad A \to \lambda.$$

(In the productions of the first type, the nonterminal symbol B is on the *left* rather than the right of the terminal symbol c.)

4.3 Examples

(a) The following grammar is right-linear:

$$S \to aS$$
$$S \to bS$$
$$S \to \lambda.$$

The set which it generates is the set $\{a,b\}^*$, as is easily verified. The word *ababb*, for example, is produced as follows:

$$S \Rightarrow aS \Rightarrow abS \Rightarrow abaS \Rightarrow ababS \Rightarrow ababbS \Rightarrow ababb.$$

(b) Here is a simple left-linear grammar:

$$S \to Aa$$
$$S \to Bb$$
$$A \to Bb$$
$$B \to Aa$$
$$A \to \lambda$$
$$B \to \lambda.$$

It produces the set of all alternating sequences of a and b.

Let G and G' be two grammars with the same terminal alphabet. Otherwise G and G' may be quite different; for example, one may be right-linear and the other left-linear, and they may have different nonterminal alphabets.

4.4 Definition. We say that G and G' are equivalent if the sets generated by them are equal, that is, if

$$L(G) = L(G').$$

4.5 Lemma. For every right-linear grammar G, there exists an equivalent right-linear grammar G' such that G' contains no productions of the form $B \to \lambda$ for $B \neq S$ and no productions of the form $A \to cS$ where S is the start symbol. Analogous properties hold for left-linear grammars.

Proof. Let us first take a typical example for the elimination of productions $A \to \lambda$:

$$\left.\begin{array}{l} S \to bB \\ B \to aB \\ B \to \lambda \end{array}\right\} P \quad \text{may be replaced by} \quad P' \left\{\begin{array}{l} S \to bB \\ B \to aB \\ S \to b \\ B \to a. \end{array}\right.$$

It is hoped that the reader sees what happened.

Next, we show how to eliminate productions $A \to cS$. Let us again take a typical example:

$$\left.\begin{array}{l} S \to bA \\ A \to aS \\ A \to a \end{array}\right\} P \quad \text{may be replaced by} \quad P' \left\{\begin{array}{l} S \to bA \\ A \to aB \\ B \to bA \\ A \to a. \end{array}\right.$$

Again, the pattern should be clear.

4.6 Proposition. For every right-linear grammar G, there exists an equivalent left-linear grammar G'.

Proof. Let us assume, which we may by Lemma 4.5, that P contains no productions of the form $B \to \lambda$ nor any of the form $A \to cS$. For example,

$$\left. \begin{array}{l} S \to bB \\ B \to bC \\ B \to aB \\ C \to a \\ B \to b \end{array} \right\} P$$

We present P now as a directed graph, as shown in Fig. 1.3. The nodes of

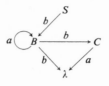

Fig. 1.3

the graph consist of the nonterminal letters, plus λ. We direct an arrow labeled c from A to B if

$$A \to cB$$

is a production. The production of a word can now be easily followed in terms of a path through the graph along the arrows, starting at S and terminating at λ. (Do this for *baaba*!) The word is produced, as is proper with right-linear grammars, from left to right. How shall we now obtain an equivalent left-linear grammar P'? Obviously P' will have to produce everything from right to left. So why don't we invert everything in sight, that is, exchange S and λ and reverse the arrows, as shown in Fig. 1.4.

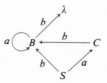

Fig. 1.4

4. GRAMMARS AND AUTOMATA

This graph corresponds to a left-linear grammar (and here we see why we wanted the special form for P):

$$\left.\begin{array}{l} S \to Bb \\ S \to Ca \\ C \to Bb \\ B \to Ba \\ B \to b \end{array}\right\} P'$$

It is clear that P' generates the same words as does P. (Just follow the path back in Fig. 1.4 for the example *baaba* that you ran through Fig. 1.3!) This procedure of obtaining P' from P is quite general, and we are done.

In view of the proposition which we have just established, the terms "right-linear language" and "left-linear language" denote the same thing. This leads us to adopt

4.7 Definition. A set of words is called a *regular language* if it is a right-linear or a left-linear language.

4.8 Problems

(a) If L_1 and L_2 are regular languages, then so is $L_1 \cup L_2$ (the set consisting of all words which are in at least one of the sets L_1 or L_2).

(b) If L_1 and L_2 are regular languages, then so is $L_1 \cdot L_2$ (the set of all words $u \cdot v$ where $u \in L_1$ and $v \in L_2$).

(c) If L is a regular language, then so is
$$L^* = \{\lambda\} \cup L \cup L \cdot L \cup L \cdot L \cdot L \cup \cdots.$$

(d) If L is a regular language, then so is L^R which consists of all reverses w^R for $w \in \Sigma$. [See (b) of Problem 1.5 for the definition of R.]

(e) Describe informally a procedure which accomplishes the following: Given a right-linear grammar P and a word w over the terminal alphabet, decide whether $w \in \Sigma(P)$ or not.

(f) The grammar (in Section 3) which generates all numbers divisible by 3 is obviously right-linear. Find an equivalent left-linear grammar.

(g) The grammar (in Section 3) which generates all true equations of addition is left-linear. Find an equivalent right-linear grammar.

(h) Devise a finite automaton for checking equations of addition (analog to the one in Section 3), but which reads the words from left to right.

(i)* Consider decimal notations $a, b, c, \ldots$ for natural numbers and let Σ be the set of all words of the form $a + b = c$ which are true arithmetic statements, for example, $312 + 14 = 326$. (Thus we consider words over the alphabet $\{0, 1, \ldots, 9, +, =\}$.) Show that Σ is not a regular language.

We now come to the second part of our task for the present section, namely to give a formal definition of the concept *finite automaton*. The graphical way of treating finite automata that we employed in Section 3 motivates our formal definition which is given by specifying a simple type of algebraic structure.[2] This approach may be less intuitive at first; but this fact is balanced by easier amenability of algebraic structures to mathematical treatment. We shall, of course, continue to use transition diagrams to specify finite automata if we can gain something by it.

4.9 Definition. Let $A = \{a, \ldots\}$ be a finite alphabet. A *finite* automaton over A consists of the following items:

(a) a finite nonempty set F, called the set of *states*;
(b) a subset D of F, called the set of *final states*;
(c) a distinguished element $s_0 \in F$, called the *start state*;
(d) a unary function $f_x: F \to F$ for each $x \in A$, $i = 1, \ldots, m$, called the *transition functions*.

For convenience, we shall present finite automata in the form of a sequence of entities, to wit

$$\langle F, D, s_0, f_a, \ldots \rangle.$$

This is the way in which algebraic structures in general are presented (for example, the monoids of Section 2). Note, however, that apart from determining the kind of mathematical entities that go into the makeup of a finite automaton, we are asking very little by the way of relations among these entities. (For example, no equations such as $f_{a_1} f_{a_2} = f_{a_2} f_{a_1}$, etc.)

---

[2] This format was proposed by Büchi (11).

4.10 Examples[3]

(a) A finite automaton may be given to us by a *transition diagram* (Fig. 1.5) such as the ones employed in Section 3. Here is how we recognize the mathematical structure of such a diagram. The *start* state is the one which we indicate by enclosing the vertex by a diamond, the final

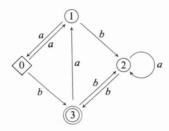

Fig. 1.5

states are circled twice, and the arrows are labeled by the nonterminal symbols as indicated. Note that $F = \{0,1,2,3\}$, $D = \{3\}$, $s_0 = 0$, and f_a, f_b are given by Table 1.1.

TABLE 1.1

s	$f_a(s)$	$f_b(s)$
0	1	3
1	0	2
2	2	3
3	1	2

(b) A finite automaton is sometimes informally presented as a set of states (with some designated states and a start state) plus a next-state function g. The function g has two variables, the first running over states, the second over the alphabet. The value of $g(s,x)$ is the next state the automaton is in if it receives an input symbol x in state s. For our example (a), the table for g would be as shown in Table 1.2.

---

[3] The first example of finite automata theory concerned the behavior of nerve nets; see McCulloch and Pitts (16).

TABLE 1.2

s	x	$g(s,x)$
0	a	1
1	a	0
2	a	2
3	a	1
0	b	3
1	b	2
2	b	3
3	b	2

TABLE 1.3
States[a]

Switch 1	Switch 2	Switch 3	
Off	Off	Off	= state A
Off	Off	On	= state B
Off	On	Off	= state C
Off	On	On	= state D
On	Off	Off	= state E
On	Off	On	= state F
On	On	Off	= state G
On	On	On	= state H

[a] *Final states:* $\{F, G\}$. *Start state:* A.

(c) In practical examples of course we have to determine first what the states, and so forth, are. Let us, for example, consider the switching circuit shown in Fig. 1.6 in which the light is on only if either switches 1 and 3 or 1 and 2 are on. Let the start state be the one in which all switches

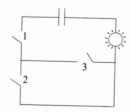

Fig. 1.6

are off. Interpret a sequence of numerals 1, 2, and 3 as a history of flipping the switches; for example, 1232 would put 1 on, 2 on, 3 on, 2 off, resulting in 1 and 3 on, 2 off. This is a finite automaton in disguise.

We now present it in the form according to Definition 4.9, where the states and the transition functions are given in Tables 1.3 and 1.4, respectively.

TABLE 1.4
Transition Functions

s	$f_1(s)$	$f_2(s)$	$f_3(s)$
A			
B	F	D	A
C			
D	H	B	C
E			
F			
G			
H			

The reader is asked to fill in Table 1.4.

Let A^* be the set of words over the finite alphabet A, and let F be a finite automaton over A. We are now in a position to define precisely what it means for F to accept a word w from A^*.

4.11 Definition

(a) If F is a finite automaton over A, we let F be the function $F: A^* \to F$ defined by

$$F(\lambda) = s_0$$
$$F(wa_i) = f_{a_i}(F(w)), \qquad i = 1, 2, \ldots, m.$$

(b) A word w in A^* is *accepted* by F if and only if $F(w) \in D$. The set of words accepted by F is denoted by $\Sigma(F)$.

In order to justify this definition, we have to convince ourselves that there does exist a unique function F satisfying the equations above. The theorem to invoke here is Proposition 1.4. To bring the defining equations into the proper form for that theorem, let us define $g: F \times A^* \times A \to F$ by $g(s, w, a_i) = f_{a_i}(s)$, $i = 1, 2, \ldots, m$. Then the equations above take on the right form.

Now that we are sure that we have formally defined a concept of acceptance, we have to make it clear to ourselves that it agrees with our intuitive description of it on the preceding pages.

4.12 Problems

(a) Describe the set of all words accepted by the automaton shown in Fig. 1.7.

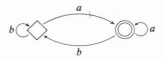

Fig. 1.7

(b) Find an automaton that accepts the set of all words

$$aabb, aabbaabb, aabbaabbaabb, \ldots$$

[i.e., all words $(a^2 b^2)^n$, $n = 1, 2, \ldots$].

(c) Let W be the set of all words

$$a^n b^n, \qquad n = 0, 1, 2, \ldots.$$

Show that there is no finite automaton which accepts W.

(d) Let F be a finite automaton and s a state of F. Let F_s be defined by $F_s(\lambda) = s$, $F_s(w \cdot x) = f_x(F_s(w))$ for all x in the alphabet A. Show that

$$F_s(u \cdot v) = F_{F_s(u)}(v)$$

for all $u, v \in A^*$.

Our main goal in this chapter is to investigate what kinds of sets are accepted by finite automata. The main result is that these are exactly the regular languages. To prove this, we shall have to construct a right-linear grammar G for each finite automaton F in such a manner that $L(G) = \Sigma(F)$. Let us do this for an example first.

4. GRAMMARS AND AUTOMATA

Consider the automaton shown in Fig. 1.8. The grammar will have three nonterminal symbols S_1, S_2, and S_3 (and, of course, the terminal symbols a and b). To find the production rules, we are led by the idea

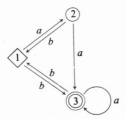

Fig. 1.8

Fig. 1.9

that the sequence of productions should parallel the sequence of moves from state to state in the automaton. Thus, S_1 (which corresponds to the start state) is the start symbol. To the arrow shown in Fig. 1.9, we let correspond a production rule.

$$S_1 \to aS_2,$$

and so on. Consider now any input sequence w to the finite automaton, and parallel the sequence of moves that the finite automaton makes on this input by a sequence of productions according to the rules given above. At the end of the last move, the production sequence will have produced the word wS_i where S_i is the state which the automaton reaches on input w. Now, w is accepted if and only if i is a final state. If w is supposed to be produced also, we should therefore add the production rule

$$S_i \to \lambda$$

for each final state i. Altogether, the grammar for the automaton shown in Fig. 1.8 is the following:

$$\begin{aligned} S_1 &\to aS_2, & S_3 &\to aS_3, \\ S_1 &\to bS_3, & S_3 &\to bS_1, \\ S_2 &\to aS_3, & S_3 &\to \lambda. \\ S_2 &\to bS_1, & & \end{aligned}$$

Let us now give a more formal statement and proof of this result.

4.13 Theorem.[4] If $A = \{a,\ldots\}$ is an alphabet, and $\Sigma \subseteq A^*$ is a set of words accepted by a finite automaton $\langle F, D, s_0, f_a, \ldots \rangle$ over A, then Σ is a right-linear language.

Proof. Let us define a right-linear grammar G as follows. As nonterminal symbols, we choose a symbol $\bar{s}$ for each state s in F (taking care that no $\bar{s}$ equals one of the symbols in A). Let $\bar{s}_0$ be the start symbol of the grammar. As productions, we choose

$$P: \begin{cases} \bar{s} \to x \cdot \overline{f_x(s)} & \text{for all} \quad s \in F \quad \text{and} \quad x \in A, \\ \bar{s} \to \lambda & \text{for all} \quad s \in D. \end{cases}$$

This grammar is obviously right-linear. We have to show that $\Sigma = L(G)$, which is done by proving $\Sigma \subseteq L(G)$ and $L(G) \subseteq \Sigma$.

Observe that for each $w \in A^*$ we have $\bar{s}_0 \stackrel{*}{\underset{P}{\Rightarrow}} w$ iff $\bar{s}_0 \stackrel{*}{\underset{P}{\Rightarrow}} w\bar{s}_j$ for some $s_j \in D$. We show presently that $\bar{s}_0 \stackrel{*}{\underset{P}{\Rightarrow}} w\bar{s}_j$ iff $s_j = F(w)$. This then implies at once that $s_0 \stackrel{*}{\underset{P}{\Rightarrow}} w$ iff $F(w) \in D$, that is, that $w \in L(G)$ iff $w \in \Sigma(F)$.

To show that $\bar{s}_0 \stackrel{*}{\underset{P}{\Rightarrow}} w\bar{s}_j$ iff $s_j = F(w)$ for all $w \in A^*$ we use Proposition 1.2. This tells us that we need only show this property for $w = \lambda$ and prove that the property for $u \cdot a_i$ follows from the property for u for each $a_i \in A$.

[4] Chomsky and Miller (12).

Now, $\bar{s}_0 \overset{*}{\underset{P}{\Rightarrow}} \lambda \bar{s}_i$ is true iff $\bar{s}_i = \bar{s}_0$; but $s_0 = F(\lambda)$, as it should. Hence the property is established for $w = \lambda$. Suppose then that the property holds for u. Assume first that $\bar{s}_0 \overset{*}{\underset{P}{\Rightarrow}} u a_i \bar{s}_j$. By induction assumption we have $\bar{s}_0 \overset{*}{\underset{P}{\Rightarrow}} u \cdot \overline{F(u)}$. Together these imply that there must be a production $\overline{F(u)} \to a_i \bar{s}_j$ in P, in other words, that $s_j = f_{a_i}(F(u)) = F(u \cdot a_i)$. Assume next that $F(u \cdot a_i) = s_j$. Hence $s_j = f_{a_i}(F(u))$ and therefore $\overline{F(u)} \to a_i \bar{s}_j$ is one of the productions in P. Using the induction hypothesis $\bar{s}_0 \overset{*}{\underset{P}{\Rightarrow}} u\overline{F(u)}$, we see that $\bar{s}_0 \overset{*}{\underset{P}{\Rightarrow}} u a_i \bar{s}_j$.

This proves the theorem. We establish the converse later as a consequence of other results.

5. REGULAR SETS

The goal of this section is to provide a mathematical characterization of regular languages. For this purpose, we first need to obtain some insight into the totality of such languages. Let us start with the limiting cases.

The empty set is a right-linear language, it is the language generated by the grammar whose only production is $S \to a_1 S$.

The set $\{\lambda\}$ consisting only of the empty word is right-linear. The appropriate grammar is $S \to \lambda$.

Each set $\{a_i\}, i = 1, \ldots, m$, is right-linear. It is generated by the grammar $S \to a_i$.

Next, we recall the results of Section 4 [(a)–(c) of Problem 4.8]: If L_1 and L_2 are regular languages, then so are

$$L_1 \cup L_2 = \{w : w \in \Sigma_1 \quad \text{or} \quad w \in \Sigma_2\},$$
$$L_1 \cdot L_2 = \{u \cdot v : u \in L_1 \quad \text{and} \quad v \in L_2\},$$
$$L_1^* = \{\lambda\} \cup L_1 \cup L_1 \cdot L_1 \cup L_1 \cdot L_1 \cup \cdots.$$

The fact that the class of regular languages is closed under $\cup$, $\cdot$, and $*$ suggests the following:

5.1 Definition.[5] Let A be an alphabet. The family of all *regular sets* over A is obtained as follows:

(a) $\emptyset$ is a regular set.
(b) $\{\lambda\}$ is a regular set.
(c) Each set $\{a_i\}$, where $a_i \in A$, is a regular set.
(d) If L_1 and L_2 are regular sets, then so are $L_1 \cup L_2$, $L_1 \cdot L_2$, and L_1^*.
(e) No other sets are regular sets.

From our previous remark about the closure properties of the class of regular languages, it follows at once that every regular set is a regular language. But the surprising result is this:

5.2 Theorem. A set $L \subseteq A^*$ is a regular set if and only if it is a regular language.

Proof. We need to show that every regular language is a regular set. Without loss of generality, we may consider only right-linear languages.

Thus, let G be a right-linear grammar consisting of n production rules. We will show that $L(G)$ is a regular set.

Let i, j, and k be numbers such that $1 \leqslant k \leqslant n+1$. Let Σ_{ij}^k be the set of words which are obtained as follows: We start with the ith rule, after that we use rules numbered less than k any number of times. But we take care that the last such rule which we apply leaves us with a word to which the jth rule would be applicable. The words obtained will therefore all end with the symbol A which is to the left of the jth production rule $A \to \ldots$. Let $\bar{\Sigma}_{ij}^k$ be the set obtained from Σ_{ij}^k by deleting these last symbols; that is, let $\bar{\Sigma}_{ij}^k = \{u : uC \in \Sigma_{ij}^k \text{ for some nonterminal symbol } C\}$.

5.3 Example. Consider the grammar

1. $S \to aA$
2. $A \to bA$
3. $A \to aS$
4. $S \to \lambda$.

---

[5] This definition is due to Kleene (14) who proved that the family of regular sets is the same as the family of all sets accepted by some automaton (this is a portion of our main theorem 6.7).

5. REGULAR SETS

The set Σ_{32}^3 contains, for example, the words produced as follows

$$A \Rightarrow aS \Rightarrow aaA \Rightarrow aabA: aabA \in \Sigma_{32}^3$$
$$A \Rightarrow aS \Rightarrow aaA: aaA \in \Sigma_{32}^3.$$

It does not contain the word produced by

$$A \Rightarrow aS: aS \notin \Sigma_{32}^3;$$

this word is, however, contained in Σ_{31}^3, for example.

Proof of Theorem 5.2 (*continued*). We show first that each set $\bar{\Sigma}_{ij}^k$ is a regular set for each i, j, and k. For $k = 1$, this is easy because the resulting sets are finite. For $k > 1$, we proceed by induction. Note that

$$\bar{\Sigma}_{ij}^k \cup \bar{\Sigma}_{ik}^k \cdot (\bar{\Sigma}_{kk}^k)^* \cdot \bar{\Sigma}_{kj}^k$$

is a regular set by definition (and induction assumption). We plan to show it equal to $\bar{\Sigma}_{ij}^{k+1}$. Clearly, if $w \in \bar{\Sigma}_{ij}^k \cup \bar{\Sigma}_{ik}^k \cdot (\bar{\Sigma}_{kk}^k)^* \cdot \bar{\Sigma}_{kj}^k$, then $w \in \bar{\Sigma}_{ij}^{k+1}$. On the other hand, consider any $w \in \bar{\Sigma}_{ij}^{k+1}$ and fix an appropriate production sequence for w (consisting of productions numbered i or at most k). If the kth production is not among them, then $w \in \bar{\Sigma}_{ij}^k$ and we are through. Otherwise $w \in \bar{\Sigma}_{ik}^k \cdot (\bar{\Sigma}_{kk}^k)^* \cdot \bar{\Sigma}_{kj}^k$ as we see after a moment's reflection.

It remains to express $L(G)$ in terms of the $\bar{\Sigma}_{ij}^k$. For this we note that a word $w \in L(G)$ is only produced if the first production is of the form $S \to \cdots$ and the last of the form $\cdots \to a_i$ or $\cdots \to \lambda$. There are only finitely many combinations of such first and last rules, each of them giving rise to a regular set. Their union makes up $L(G)$. Hence $L(G)$ is a regular set.

5.4 Example. Determine the set generated by the grammar

$$P: \begin{cases} S \to aA & (1) \\ A \to bA & (2) \\ A \to a. & (3) \end{cases}$$

Solution. $\Sigma(P) = \bar{\Sigma}_{13}^4 \cdot a$.

$\bar{\Sigma}_{13}^4 = \bar{\Sigma}_{13}^3 \cup \bar{\Sigma}_{13}^3 \cdot (\bar{\Sigma}_{33}^3)^* \cdot \bar{\Sigma}_{33}^3 = \bar{\Sigma}_{13}^3 \cup \bar{\Sigma}_{13}^3 \cdot \phi^* \cdot \phi = \bar{\Sigma}_{13}^3$

$\bar{\Sigma}_{13}^3 = \bar{\Sigma}_{13}^2 \cup \bar{\Sigma}_{12}^2 \cdot (\bar{\Sigma}_{22}^2)^* \cdot \bar{\Sigma}_{23}^2$

$\bar{\Sigma}_{13}^2 = \bar{\Sigma}_{13}^1 \cup \bar{\Sigma}_{11}^1 \cdot (\bar{\Sigma}_{11}^1)^* \cdot \bar{\Sigma}_{13}^1 = a \cup \phi \cdot (\cdots) = a$

$\bar{\Sigma}_{12}^2 = \bar{\Sigma}_{12}^1 \cup \bar{\Sigma}_{11}^1 \cdot (\bar{\Sigma}_{11}^1)^* \cdot \bar{\Sigma}_{12}^1 = a \cup \phi \cdot (\cdots) = a$

$\bar{\Sigma}_{22}^2 = \bar{\Sigma}_{22}^1 \cup \bar{\Sigma}_{21}^1 \cdot (\bar{\Sigma}_{11}^1)^* \cdot \bar{\Sigma}_{12}^1 = b \cup \phi \cdot (\cdots) = b$

$\bar{\Sigma}_{23}^2 = \bar{\Sigma}_{23}^1 \cup \bar{\Sigma}_{21}^1 \cdot (\bar{\Sigma}_{11}^1)^* \cdot \bar{\Sigma}_{13}^1 = b \cup \phi \cdot (\cdots) = b$

$\bar{\Sigma}_{13}^3 = a \cup a \cdot b^* b$

$\bar{\Sigma}_{13}^4 = a \cup a \cdot b^* b$

$\boldsymbol{\Sigma(P)} = (a \cup ab^*b) \cdot a = aa \cup ab^*ba = \boldsymbol{ab^*a}$.

Finite automata are a source of regular sets, since by Theorem 4.13 each set accepted by a finite automaton is a regular language and hence a regular set.

5.5 Corollary. The set accepted by a finite automaton is a regular set.

It is, of course, possible to see this directly, without appeal to Theorem 4.13. The following example indicates the general procedure.

5.6 Example. Consider the automaton F given by the diagram

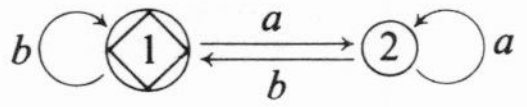

Fig. 1.10

shown in Fig. 1.10. Let E_{ij}^k be the set of words such that, if the automaton is started in state i on w, it goes exclusively through states numbered less than k and finishes in state j. Let us write a, b, ..., instead of $\{a\}$, $\{b\}$, ...

5. REGULAR SETS

$$\Sigma(F) = E^3_{11} = E^2_{11} \cup E^2_{12} \cdot (E^2_{22})^* \cdot E^2_{21} = b^* \cup (b^*a)(E^2_{22})^* \cdot E^2_{21}$$
$$E^2_{22} = E^1_{22} \cup E^1_{21} \cdot (E^1_{11})^* \cdot E^1_{12} = a \cup b \cdot b^*a = b^*a$$
$$E^2_{21} = E^1_{21} \cup E^1_{21} \cdot (E^1_{11})^* \cdot E^1_{11} = b \cup b \cdot b^* \cdot b = bb^*$$
$$\boldsymbol{\Sigma(F) = b^* \cup b^*a(b^*a)^*bb^*.}$$

5.7 Proposition. If L_1 and L_2 are regular sets, then so is their intersection $L_1 \cap L_2 = \{w : w \in L_1 \text{ and } w \in L_2\}$.

The proof of this proposition follows most easily from work below (see Theorem 6.10, whose proof, of course, does not depend on the present proposition). For now, let us consider some applications of this result.

We can make use of closure of the family of regular sets under intersection to show that certain sets L are not regular. All we have to show is that $L \cap M$ is not regular for some appropriately chosen regular set M.

5.8 Example.[6] Let $A = B = \{0, 1\}$, and let us interpret words over $\{0, 1\}$ as notations for natural numbers to the base 2. Thus the following are the first few numbers in notation to base 2:

$$0, 1, 10, 11, 100, 101, 110, 111, \ldots$$

for

$$0, 1, 2, 3, 4, 5, 6, 7, \ldots,$$

respectively. Let L be the set of numerals (to base 2) which are squares,

$$L = \{0, 1, 100, 1001, \ldots\}.$$

To show that L is not regular it is sufficient to exhibit a regular set M such that $L \cap M$ is not regular.

Let M be the set of all numerals (to base 2) of the form $(2^n - 1) \cdot 2^l + 1$, where $n > 0$, $l > 1$, and $n + l$ is even.

[6] Ritchie (20).

(a) M is regular. Namely, M is accepted by the automaton (with start state s_0) shown in Fig. 1.11. Namely,

$F(w) = s_1$ iff $w = 11\cdots 1 = 1^{(n)}$, n odd;

$F(w) = s_2$ iff $w = 1^{(n)}$, n even, $n > 0$;

$F(w) = s_3$ iff $w = 1^{(n)}0^{(m)}$, $n + m$ even; $n, m > 0$;

$F(w) = s_4$ iff $w = 1^{(n)}0^{(m)}$, $m + m$ odd, $n, m > 0$;

$F(w) = s_5$ iff $w = 1^{(n)}0^{(m)}1$, $n + m$ odd, $n, m > 0$.

It remains to note that words w with $F(w) = s_5$ are exactly the notations for numbers $(2^n - 1) \cdot 2^l + 1$, to wit $1^{(n)}0^{(l-1)}1$.

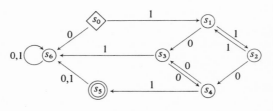

Fig. 1.11

(b) Let us consider $K = L \cap \Gamma$, and show that K is the set of numerals for numbers of the form $(2^{n+1} - 1)^2$, $n > 0$. Each numeral $(2^{n+1} - 1)^2$ is in K. Namely, observe that

$$(2^{n+1} - 1)^2 = 2^{(n+1)2} - 2^{n+2} + 1 = (2^n - 1) \cdot 2^{n+2} + 1.$$

Conversely, suppose that $x = y^2 = (2^n - 1) \cdot 2^l + 1$, $n > 0$, $l > 1$, $n + l$ even. We shall show that $y = 2^{(n+l)/2} - 1$ and $l = n + 2$. Now $y + 1$ and $y - 1$ are both even, hence not both divisible by 4; furthermore

$$(y + 1)(y - 1) = (2^n - 1) \cdot 2^l$$

and therefore either $y + 1$ or $y - 1$ (but not both) is divisible by 2^{l-1}; $y + 1 = 2^{l-1} \cdot g$, g odd. Hence

$$(2^n - 1) \cdot 2^l = (y + 1) \cdot (y - 1) = (2^{l-1} \cdot g)(2^{l-1} \cdot g \pm 2) = 2^l \cdot g \cdot (2^{l-2} \cdot g \pm 2).$$

Therefore $n \geq l - 2$ if $y^2 = (2^n - 1) \cdot 2^l + 1$ has a solution at all. Now, if

$n > l - 2$ and $n + 1$ is even, then this equation has no integral solution. Namely,

$$(2^{(l+n)/2} - 1)^2 = 2^{l+n} - 2^{(l+n)/2+1} + 1 = 2^{n+l} - 2^{(l+n+2)/2} + 1$$

which is less than $2^{n+l} - 2^l + 1$, while $2^{n+l} - 2^l + 1 < (2^{(l+n)/2})^2 = 2^{n+l}$. On the other hand, if $n = l - 2$, then $(2^{(l+n)/2} - 1)$ is a solution as can easily be verified.

(c) Finally, we have to show that K is not regular. Recall that Δ is the set of all numerals for the numbers $(2^{n+1} - 1)^2, n > 0$, which is the set of numbers $(2^n - 1) \cdot 2^{n+2} + 1$. Hence K consists of all numerals $1^{(n)} 0^{(n+1)} 1$. This set is not regular as can be seen from (c) of Problem 4.12, slightly modified.

Since every regular set of words over an alphabet A can be obtained from the empty set and the one-element sets $\{\lambda\}$, $\{a_i\}$, $a_i \in A$, by the operations ∪, ·, and *, it is reasonable to introduce *notation* for regular sets which reflect this state of affairs.

5.9 Definition

(a) Let A be a finite alphabet and let it be augmented by symbols ∅, λ, ∪, ·, *, and parentheses. Then the *regular expressions* over A are obtained as follows:

(i) ∅ and λ are regular expressions.
(ii) If $x \in A$, then x is a regular expression.
(iii) If ρ and σ are regular expressions, then so are $(\rho \cup \sigma)$, $(\rho \cdot \sigma)$, and $(\rho)^*$.
(iv) No other sequences of symbols from the augmented alphabet are regular expressions.

(b) Each regular expression ρ over A *denotes* a regular set over A in the obvious fashion.

(i) ∅ denotes the empty set.
(ii) λ denotes the set $\{\lambda\}$ consisting of the empty word only.
(iii) x denotes the set $\{x\}$ for $x \in A$.
(iv) If ρ denotes A and σ denotes B, then $(\rho \cup \sigma)$ denotes $A \cup B$, $(\rho \cdot \sigma)$ denotes $A \cdot B$, and $(\rho)^*$ denotes A^*.

If ρ denotes A, we usually indicate this by writing $|\rho| = A$.

5.10 Problems

(a) Find a regular expression ρ such that $|\rho|$ is the set of words accepted by the finite automaton shown in Fig. 1.12 where 1 is the only designated state.

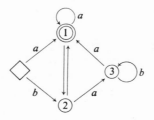

Fig. 1.12

(b) Prove directly (i.e., without using right-linear grammars) that the set of words accepted by a finite automaton is regular.

(c) Show that $|\lambda| = |\rho|$ for some regular expression ρ which does not contain the symbol λ. Thus, strictly speaking, we would not need the symbol λ in our notation system.

(d) If A, B, and C are regular sets, then the following equations are true. Prove them.

$$\phi \cdot A = \phi \qquad A \cup B = B \cup A$$
$$A \cup A = A \qquad A \cup (B \cup C) = (A \cup B) \cup C$$
$$\phi^* \cdot A = A \qquad A \cdot (B \cdot C) = (A \cdot B) \cdot C$$
$$A \cup \phi = A \qquad A \cdot (B \cup C) = (A \cdot B) \cup (A \cdot C)$$
$$A^* = \phi^* \cup A^* \cdot A \qquad (A \cup B) \cdot C = (A \cdot C) \cup (B \cdot C)$$
$$A^* = (\phi^* \cup A)^*$$

[*Remark*: The 11 identities above constitute an axiom system for an "algebra of regular sets" proposed by Salomaa (21).]

(e) Let A be a regular set. Consider perm(A), the set of all words obtained from words in A by permuting the symbols in some fashion. For example, if $xyzzx \in A$, then the following words would be in perm(A): $xxzzy$, $yxzxz$, $zxzxy$, $zxyxz$, ... Is perm(A) a regular set whenever A is?

6. BUILDING AN ACCEPTOR

We have shown so far that the set of words accepted by a finite automaton is a right-linear language. The purpose of the present section is to prove the converse. Let us first consider an example.

Example. Consider the right-linear grammar

$$\left.\begin{array}{l} S \to aS \\ S \to bB \\ B \to cC \\ C \to aS \\ S \to b \end{array}\right\} \begin{array}{l} \text{terminal symbols} \quad a, b, c \\ \text{nonterminal symbols} \quad S, B, C. \end{array}$$

We can visualize productions in this grammar in a diagrammatic form (Fig. 1.13) akin to the transition diagram of a finite automaton. Obviously a word $w \in \{a,b,c\}^*$ is produced by the grammar iff there is a path through the diagram starting at (S) and terminating at (λ) such that the labels of the arrows that are successively traveled through spell the word w. For example, *aaabcaabcab* is accepted.

The difference between the diagram of Fig. 1.13 and a diagram for a finite automaton consists in two things.

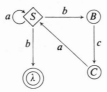

Fig. 1.13

First. The node (B), for example, has only one arrow emanating from it, labeled "c." This makes the diagram *incomplete*. This is easily corrected, without changing the set accepted, by introducing an additional node (T) with the connections shown in Fig. 1.14.

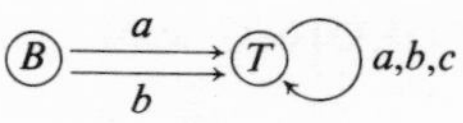

$B \underset{b}{\overset{a}{\rightrightarrows}} T \circlearrowright a,b,c$

Fig. 1.14

Second. The node $\langle S \rangle$, for example, has two arrows labeled "b" emanating from it. This gives us a free choice in starting on a path if the word begins with b. This makes the diagram *nondeterministic*. This is not so easily corrected. However, nondeterministic finite automata are a helpful intermediate step in the process of finding an automaton that accepts a given right-linear language.[7]

6.1 Definition

(a) A *nondeterministic* finite automaton over an alphabet $A = \{a_1,\ldots,a_m\}$ is a structure

$$\langle F, D, s_0, F_{a_1},\ldots, F_{a_m}\rangle$$

where F is a finite set; the set of states D is a subset of F (the set of final states); $s_0 \in F$; F_{a_i}, $i = 1, \ldots, m$, are binary relations on F (that is, $F_{a_i} \subseteq F \times F$) called the transition relations.

(b) Let w be a word over A; the set of *possible states after input w* is defined recursively by

$$P(\lambda) = \{s_0\}$$

$$P(w \cdot a_i) = \{s \in F: \text{ there exists } s' \in P(w) \text{ such that } \langle s', s \rangle \in F_{a_i}\}.$$

(c) We say that $w \in A^*$ is *accepted* by the nondeterministic automaton F if one of the possible states after input w is an element of D; that is, iff $P(w) \cap D \neq \emptyset$. We denote by $\Sigma(F)$ the set of words accepted by F.

In the example above we have $F = \{S, B, C, \lambda\}$, $s_0 = S$, $D = \{\lambda\}$, $F_a = \{\langle S, S\rangle, \langle C, S\rangle\}$, $F_b = \{\langle S, B\rangle, \langle S, \lambda\rangle\}$, $F_c = \{\langle B, C\rangle\}$. This is a nondeterministic finite automaton which accepts the language produced by the grammar of our example.

---

[7] This was first employed by Rabin and Scott (19).

6. BUILDING AN ACCEPTOR

6.2 Remark. The term "nondeterministic" is somewhat too strong, for we easily recognize that each "deterministic" finite automaton is nondeterministic in the sense discussed above. Namely, if

$$f_a: F \to F$$

is a transition function, associate to it the transition relation

$$F_a = \{\langle s, f_a(s) \rangle : s \in S\}.$$

With this correspondence, the deterministic finite automaton

$$\langle F, D, s_0, f_{a_1}, \ldots \rangle$$

is identical with the nondeterministic finite automaton

$$\langle F, D, s_0, F_{a_1}, \ldots \rangle.$$

In general, however, the sets F_a of nondeterministic automata are not restricted to such sets of pairs arising as ("graphs" of) functions; this is what makes them nondeterministic.

The construction of a nondeterministic automaton employed above suggests a general procedure.

6.3 Theorem. For every right-linear language, there exists a nondeterministic finite automaton which accepts it.

Proof. We choose lower-case letters to denote elements of the terminal alphabet and capital letters to denote elements of the nonterminal alphabet of the given right-linear grammar G. Let S be the start symbol.

Consider the following nondeterministic finite automaton

$$\langle F, D, s_0, F_{a_1}, \ldots, F_{a_m} \rangle:$$

$s_0 = S$
$F = \{\lambda\} \cup \{R : R \text{ is a nonterminal symbol of } G\}$
$D = \{\lambda\} \cup \{R : R \to \lambda \text{ is a production of } G\}$
$F_{a_i} = \{\langle R, T \rangle : R \to a_i T \text{ is a production of } G\}$
$\cup \{\langle R, \lambda \rangle : R \to a_i \text{ is a production of } G\}.$

We first prove the following fact: For all nonterminal symbols R of the grammar G and all $w \in A^*$ we have $S \overset{*}{\Rightarrow} wR$ iff $R \in P(w)$.

For $w = \lambda$ this is trivial: $S \overset{*}{\Rightarrow} R$ iff $R = S$ iff $R \in P(\lambda) = \{S\}$. Assume the statement for u; and suppose $S \overset{*}{\Rightarrow} ua_i T$. Then for some R we must have $S \overset{*}{\Rightarrow} uR \Rightarrow ua_i T$, and $R \to a_i T$ must be a production in G. Thus $\langle R, T \rangle \in F_{a_i}$, $R \in P(u)$, hence $T \in P(ua_i)$. Conversely, suppose that $T \in P(ua_i)$. By definition of P there exists R such that $R \in P(u)$ and $\langle R, T \rangle \in F_{a_i}$; hence $R \to a_i T$ is a production in G. Since $R \in P(u)$ by assumption, it follows that $S \overset{*}{\Rightarrow} uR \Rightarrow ua_i T$.

Suppose now that $S \overset{*}{\Rightarrow} w$ where $w \in A^*$. We have the following cases.

First case: $S \overset{*}{\Rightarrow} wR \Rightarrow w$; that is, $R \to \lambda$ is a production. Then $R \in D$ by definition, $R \in P(w)$ by the above. Hence $P(w) \cap D \neq \varnothing$ and w is accepted by the automaton.

Second case: $S \overset{*}{\Rightarrow} uR \Rightarrow ua_i = w$; that is, $R \to a_i$ is a production. Then $\langle R, \lambda \rangle \in F_{a_i}$, $R \in P(u)$, $\lambda \in P(ua_i) = P(w)$. Hence $P(w) \cap D \neq \varnothing$ since both contain λ.

Conversely, suppose that $P(w) \cap D \neq \varnothing$, say $T \in P(w) \cap D$.

First case: $T \neq \lambda$. Then $S \overset{*}{\Rightarrow} wT$ by the fact proven above. Since $T \in D$ we must have that $T \to \lambda$ is a production in G, hence $S \overset{*}{\Rightarrow} w$.

Second case: $T = \lambda$. Thus $\lambda \in P(w)$, and therefore w cannot be the empty word, $w = ua_i$. Hence there exists $R \in P(u)$ with $\langle R, \lambda \rangle \in F_{a_i}$. It follows that $S \overset{*}{\Rightarrow} uR$ and that $R \to a_i$ is a production. Hence $S \overset{*}{\Rightarrow} ua_i = w$.

6.4 Theorem. For every nondeterministic finite automaton, there exists a (deterministic) finite automaton which accepts the same set of words.

Proof. Conceptually, the proof is quite straightforward. Suppose we are given a word w and want to see whether it is accepted by some nondeterministic automaton $\langle F, D, s_0, F_{a_1}, \ldots, F_{a_n} \rangle$. Imagine that we have a large supply of copies of this automaton and let us watch these little machines simultaneously chugging away at the word w. At the start, all

6. BUILDING AN ACCEPTOR

machines are in state s_0. But after this they may diverge. Let us note down, after each move, what the states of the individual machines are. We do not even care in what order these states are written down, it is only the *set* of states attained that we wish to consider. If w has length n, then all machines stop after n moves and we get a final collection of states attained. The word w is accepted if there is a distinguished state in this collection.

Can we predict the set of states which will be reached by the collection of automata if we know the present set of states and the input symbol? We clearly can, just by looking at the transition relations. Indeed, we can build a finite automaton that will do the predicting for us. We are after

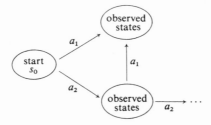

Fig. 1.15

the finite automaton depicted in Fig. 1.15. If M is a set of observed states and a_i is a symbol, then the set M' of observed states after input a_i is obtained by

$s' \in M'$ iff there exists $s \in M$ such that $\langle s, s' \rangle \in F_{a_i}$.

The passage from M to M', given a_i, is uniquely determined. In fact, we can define a function f_{a_i} ranging over the set of all subsets of F such that

$$f_{a_i}(M) = \{s' : \langle s, s' \rangle \in F_{a_i} \text{ for some } s \in M\}.$$

Let F^s be the set of all subsets of F (where F is the set of states of the given nondeterministic automaton) and let D^s be the collection of those subsets of F which contain at least one element of D (where D is the set of distinguished states of the nondeterministic automaton). Consider the finite automaton

$$\langle F^s, D^s, \{s_0\}, f_{a_1}, \ldots, f_{a_m}\rangle.$$

Then $\{s_0\}$, a one-element subset of F, is clearly an element of F^s; we choose it as the starting state of F^s.

From the way that we arrived at the construction of F^s, namely by watching performances of many copies of F, we expect that the set accepted by F is the same as that accepted by F^s. Let us now give a more formal proof of this fact.

Let W be the set of words w over A such that $F^s(w) = P(w)$. We show first that $W = A^*$. By Section 1 we know that this reduces to proving $\lambda \in W$ and $w \in W$ implies $w \cdot a_i \in W$ for all $a_i \in A$.

Note that $\lambda \in W$ since $F^s(w) = \{s_0\} = P(\lambda)$ by definition.

Assume $F^s(w) = P(w)$. Then

$$\begin{aligned}F^s(w \cdot a_i) &= f_{a_i}(F^s(w)) = f_{a_i}(P(w))\\ &= \{s': \langle s, s'\rangle \in F_{a_i} \text{ for some } s \in P(w)\} = P(w \cdot a_i)\end{aligned}$$

by definition of F^s, assumption, definition of f_{a_i}, and definition of P, respectively.

In particular, if w is accepted by the deterministic automaton, that is, if $F^s(w) \in D^s$, then $P(w) \cap D \neq \varnothing$ and w is also accepted by the nondeterministic automaton. Similarly, if w is accepted by the nondeterministic automaton, that is, if $P(w) \cap D \neq \varnothing$, then $P(w) = F^s(w) \in D^s$, and hence w is also accepted by the deterministic automaton.

6.5 Remark. In the construction of the set F^s of states of the deterministic automaton, we may restrict ourselves to those sets of states which can actually be reached from $\{s_0\}$ by repeatedly applying the operations f_{a_i}. All others will not contribute to the recognizing capability of the automaton constructed. This will, in general, result in a smaller automaton (and shorter construction).

6.6 Example. Find a deterministic automaton which accepts the same set as the nondeterministic automaton shown in Fig. 1.16.

6. BUILDING AN ACCEPTOR

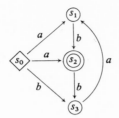

Fig. 1.16

Solution

$$f_a(\{s_0\}) = \{s_1, s_2\}$$
$$f_b(\{s_0\}) = \{s_3\}$$
$$f_a(\{s_1, s_2\}) = \varnothing$$
$$f_b(\{s_1, s_2\}) = \{s_2, s_3\}$$
$$f_a(\{s_3\}) = \{s_1\}$$
$$f_b(\{s_3\}) = \varnothing$$
$$f_a(\varnothing) = \varnothing$$
$$f_b(\varnothing) = \varnothing$$
$$f_a(\{s_2, s_3\}) = \{s_1\}$$
$$f_b(\{s_2, s_3\}) = \{s_3\}$$
$$f_a(\{s_1\}) = \varnothing$$
$$f_b(\{s_1\}) = \{s_2\}$$
$$f_a(\{s_2\}) = \varnothing$$
$$f_b(\{s_2\}) = \{s_3\}$$
$$F = \{\varnothing, \{s_0\}, \{s_1\}, \{s_2\}, \{s_3\}, \{s_1, s_2\}, \{s_2, s_3\}\}$$
$$D = \{\{s_2\}, \{s_1, s_2\}, \{s_2, s_3\}\}$$
$$\langle F, D, \{s_0\}, f_a, f_b \rangle.$$

Let us now sum up our results.

1. INTRODUCTION TO FINITE AUTOMATA

6.7 Main Theorem. For any set W or words over an alphabet A, the following statements are equivalent:

(i) W is produced by a regular grammar.
(ii) W is a regular set.
(iii) W is accepted by a deterministic finite automaton.
(iv) W is accepted by a nondeterministic finite automaton.

Proof. We need only collect the relevant propositions as in the diagram below.

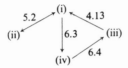

Some of the interconnections that can now be stated, such as (ii) ↔ (iii), are of independent interest. They can, of course, be obtained directly [as was indicated for (iii) → (ii) in Example 5.5]. The connection (ii) → (iv) is illustrated by Example 6.8 below. The reader may amuse himself by constructing direct proofs for some of the other connections.

6.8 Example. Find a nondeterministic automaton that accepts the set

$$(ab^* \vee ac) \cdot b \vee (a \vee b)(cca^*b)^*.$$

Solution. The solution is shown diagrammatically in Figs. 1.17–1.25 where (∗) represents start states, and ◯ , designated states.

Fig. 1.17. ab^*. *Fig. 1.18.* $a \vee b$.

6. BUILDING AN ACCEPTOR

Fig. 1.19. ac.

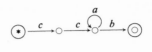

Fig. 1.20. cca^*b.

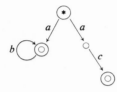

Fig. 1.21. $ab^* \lor ac$.

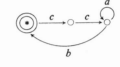

Fig. 1.22. $(cca^*b)^*$.

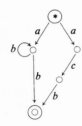

Fig. 1.23. $(ab^* \lor ac)b$.

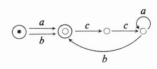

Fig. 1.24. $(a \lor b)(cca^*b)^*$.

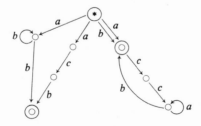

Fig. 1.25. $(ab^* \lor ac)b \lor (a \lor b)(cca^*b)^*$.

The fact that every regular set is the set accepted by a finite automaton can be used to give easy proofs of some general results about regular sets.

6.9 Theorem. If $W \subseteq A^*$ is regular, then so is $A^* - W$, the set consisting of all words not in W.

Proof. Suppose that W is accepted by the finite automaton

$$\langle F, D, s_0, f_a, \ldots \rangle.$$

Let D' be the set of all states $s \in F$ which are not in D, and consider the finite automaton

$$\langle F, D', s_0, f_a, \ldots \rangle$$

obtained by changing the set of designated states from D to D'. It is quite clear that this automaton accepts $A^* - W$, which is, therefore, regular.

6.10 Theorem. If W_1 and W_2 are regular subsets of A^*, then their intersection $W_1 \cap W_2$ is a regular set. (This is Proposition 5.7, whose proof was delayed to this point.)

Proof. We may assume that we are given two automata

$$\langle F^{(1)}, D^{(1)}, s^{(1)}, f_a^{(1)}, \ldots \rangle,$$
$$\langle F^{(2)}, D^{(2)}, s_0^{(2)}, f_a^{(2)}, \ldots \rangle,$$

accepting W_1 and W_2, respectively. Consider the direct product

$$\langle F, D, s_0, f_a, \ldots \rangle$$

of these automata, defined as follows:

$$F = \{\langle s, s' \rangle : s \in F^{(1)}, s' \in F^{(2)}\},$$
$$D = \{\langle s, s' \rangle : s \in D^{(1)}, s' \in D^{(2)}\}$$
$$s_0 = \langle s_0^{(1)}, s_0^{(2)} \rangle,$$
$$f_a(\langle s, s' \rangle) = \langle f_a^{(1)}(s), f_a^{(2)}(s') \rangle \quad \text{for all} \quad s \in F^{(1)}, \quad s' \in F^{(2)}, \quad a \in A.$$

The reason why this automaton accepts $W_1 \cap W_2$ is obvious. What it actually represents is the two given automata working in parallel. If both automata accept a word, then and only then will the combined automaton accept it.

6.11 Problems

(a) Let W be the regular set denoted by $(a \lor b^*ab)^*$. Find a regular expression denoting the set of all words not in W.

(b) Consider an alphabet consisting only of the symbol 1. Let Σ_k be the set of all sequences of 1's whose length is a multiple of k where k is any positive integer. Show that Σ_k is a regular set, and find automata which accept Σ_2, Σ_3, and Σ_6. Since $\Sigma_6 = \Sigma_2 \cap \Sigma_3$, construct an automaton for Σ_6 by using the procedure in the proof of Theorem 6.10.

7. FINITE TRANSDUCERS, MINIMALIZATION

We continue to use the words "computer and finite computer" in an informal way. The concept of finite automaton arose by making formally clear what we mean when we say that a finite computer is used as an acceptor. The notion of *finite transducer* will arise by formalizing the concept of using a finite computer as a computer; that is, if the computer is allowed to produce an output. Finite automata do not have any output. Or do they? In a sense, the automata do, indeed, have an output. They *tell* us, by their final state being distinguished or not, whether a word is accepted. Thus we may regard any finite automaton as equipped with an output device, which emits the symbol 1 if the present state is designated, or the symbol 0 if it is not.

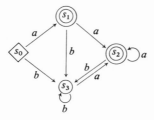

Fig. 1.26

We may indicate this in the diagram which describes the computer as follows: Instead of providing the nodes corresponding to designated states with a second circle, for example, as shown in Fig. 1.26, we represent this machine by the diagram shown in Fig. 1.27. We thus distinguish between an *input alphabet*, here $\{a,b\}$, and an *output alphabet*, here $\{0,1\}$. These need not be different.

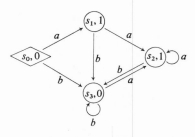

Fig. 1.27

7.1 Example. Consider the finite computer with input alphabet and output alphabet $\{0, 1, \square\}$ ($\square$ for "blank") given by the diagram shown in Fig. 1.28. If this computer is fed the word $1011011\square\square$, the sequence of symbols emitted by the states as they are being reached would

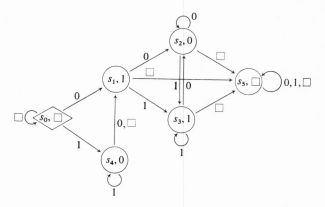

Fig. 1.28

be □0111011□□. For input 111□□□, we would get output □0001□□. Let us consider these words as notations for natural numbers in binary notation *written in reverse order*. After turning the words around, we see that

$$\square\square 1101101 \text{ transforms to } \square\square 1101110\square,$$
$$\square\square\square 111 \text{ transforms to } \square\square 1000\square.$$

From these examples, it appears that the finite machine depicted in Fig. 1.28 computes the successor function $f(x) = x + 1$. (Does it?)

7.2 Definition. A *finite transducer* with input alphabet $A = \{a_1, \ldots, a_m\}$ and *output alphabet* $B = \{b_1, \ldots, b_n\}$ is a structure

$$\langle F, s_0, f_{a_1}, \ldots, f_{a_m} \rangle$$

together with an *output assignment* $\Phi: F \to B$; $s_0 \in F$ is called the *start state*, $f_{a_1}, \ldots,$ are called the *transition function*.

(We shall from now on use the words "finite computer" and "finite transducer" interchangeably.)

We have learned, in Example 7.1, to distinguish between the actual output produced for an actual input (both contain blank symbols) and the function computed by the machine (blanks are disregarded). The former notion is the simpler one. We give a formal definition for both below.

7.3 Definition
(a) If Φ is the output assignment of a finite transducer

$$\langle F, s_0, f_{a_1}, \ldots, f_{a_m} \rangle$$

over the input alphabet $A = \{a_1, \ldots, a_n\}$, we let the *output function* $\Phi: A^* \to B^*$ be defined recursively by

$$\Phi(\lambda) = \Phi(s_0)$$
$$\Phi(w \cdot a_i) = \Phi(w) \cdot \Phi(F(w \cdot a_i)) \quad \text{for all} \quad w \in A^*, \quad a_i \in A.$$

(b) Assume that A and B contain a blank symbol □, and let A_0, B_0

result from A, B by removing the blanks. A function $f: A_0^* \to B_0^*$ is said to be *computable by a finite transducer* with input alphabet A, output alphabet B, and output assignment Φ, if, for each $w \in A_0$, we have

$$\Phi(w\square\square \cdots \square) = \square f(w)\square\square \cdots \square$$

for all input sequences $w\square\square \cdots \square\square$ which have sufficiently many blanks on the right.

7.4 Problems

(a) Show that if Σ is a regular set over A and Φ is the output function of a finite computer, then $\{w: \Phi(w) \in \Sigma\}$ is a regular set.

(b) Under the same assumptions, show that $\{\Phi(w): w \in A^*\}$ is a regular set.

(c) Construct a finite computer which adds any two numbers in binary notation as follows: $10110 + 11011$ is understood as a sequence consisting of pairs of digits at corresponding positions, to wit

$$\binom{0}{1} \binom{1}{1} \binom{1}{0} \binom{0}{1} \binom{1}{1}.$$

Note that we have taken the least significant digits first. (In case the two summands have unequal length, fill up by 0's!) Add a blank symbol and obtain the input alphabet $\{\binom{0}{0}, \binom{0}{1}, \binom{1}{0}, \binom{1}{1}, \square\}$. As output alphabet, choose $\{0, 1, \square\}$.

(d) Same problem as (c) but for multiplication of an input by 2.

(e)* Show that multiplication cannot be obtained in this manner.

Let us now fix on a particular input alphabet A and output alphabet B.

It is quite obvious that any function $\Psi: A^* \to B^*$, if it is the output function of a finite computer at all, can then be computed by very different kinds of finite computers, smaller and bigger ones. The goal of this section is to show that there exists a smallest finite computer to do the job. (This fact in itself is of course trivial. If K is the set of numbers of states of finite machines computing Ψ, then K has a smallest element.) What actually is proved is that there is a unique smallest computer.[8]

[8] This was first proved by Huffman and Moore [see (7)]; our proof follows ideas of Myhill and of Nerode (17).

7.5 Definition
(a) Two finite computers

$$(\langle F, s_0, f_{a_1}, \ldots, f_{a_m}\rangle, \Phi) \quad \text{and} \quad (\langle G, t_0, g_{a_1}, \ldots, g_{a_m}\rangle, \Psi)$$

are *equivalent* if

$$\Phi(w) = \Psi(w)$$

for all $w \in A^*$.

(b) If $s \in F$ and $t \in G$, then s *is equivalent to* t if the computers $(\langle F, s, f_{a_1}, \ldots\rangle, \Phi)$ and $(\langle G, t, g_{a_1}, \ldots\rangle, \Psi)$ are equivalent.

Thus, two computers are equivalent if they "do the same thing"; two states are equivalent if the computers do the same thing when these states are chosen as start states of the respective computers. (Of course, we are not excluding the possibility that both computers are the same.)

Let $(\langle F, s_0, f_{a_1}, \ldots, f_{a_m}\rangle, \Phi)$ and $(\langle G, t_0, g_{a_1}, \ldots, g_{a_m}\rangle, \Psi)$ be finite computers, and let $s \in F$ and $t \in G$.

7.6 Lemma. If s is equivalent to t, then $f_{a_i}(s)$ is equivalent to $g_{a_i}(t)$ for each $a_i \in A = \{a_1, \ldots, a_m\}$.

Proof. Let Φ_s be defined by the recursion equations $\Phi_s(\lambda) = \Phi(s)$, $\Phi_s(w \cdot a_i) = \Phi_s(w) \cdot \Phi(F_s(w \cdot a_i))$, where F_s in turn is defined as in Section 4, namely by $F_s(\lambda) = s$ and $F_s(w \cdot a_i) = f_{a_i}(F_s(w))$. Similar definitions give us G_t and Ψ_t. Now note that $\Phi_{f_{a_i}(s)}(w)$ is the same as $\Phi_s(a_i w)$ when we delete the first symbol of $\Phi_s(a_i w)$; we write $\Phi_{f_{a_i}(s)}(w) = d(\Phi_s(a_i w))$, using the deletion function d, defined by $d(\lambda) = \lambda$, $d(a_j w) = w$. Similarly $\Psi_{g_{a_i}(t)}(w) = d(\Psi_t(a_i w))$. By assumption, we have $\Phi_s(a_i w) = \Psi_t(a_i w)$ for all a_i and w; hence

$$\Phi_{f_{a_i}(s)}(w) = d(\Phi_s(a_i w)) = d(\Psi_t(a_i w)) = \Psi_{g_{a_i}(t)}(w) \qquad \text{for all} \quad w,$$

that is $f_{a_i}(s)$ is equivalent to $g_{a_i}(t)$.

Let us now consider a particular finite computer $(\langle F, s_0, f_{a_1}, \ldots, f_{a_m}\rangle, \Phi)$. The notion of equivalence between states applies in particular to the states of this computer; it produces a relation $\equiv$ on F. We write $s_i \equiv s_j$

if the states s_i and s_j are equivalent. Let us note the following properties of the relation $\equiv$.

(i) $s_i \equiv s_i$: "$\equiv$ is reflexive."
(ii) If $s_i \equiv s_j$ and $s_j \equiv s_k$, then $s_i \equiv s_k$: "$\equiv$ is transitive."
(iii) If $s_i \equiv s_j$, then $s_j \equiv s_i$: "$\equiv$ is symmetric."
(iv) If $s_i \equiv s_j$, then $f_{a_k}(s_i) \equiv f_{a_k}(s_j)$.

Properties (i)–(iii) are obvious from the definition; property (iv) is established in Lemma 7.6. The reason that we enumerate these four properties is that we want to generalize the situation and speak about binary relations on F, which, in general, share these properties with the particular relation $\equiv$.

7.7 Definition
(a) A binary relation $\equiv$ on F is a *congruence relation* if it satisfies properties (i)–(iv) above.
(b) A *partition* of F is a set of nonempty subsets of F such that no two such sets have a common element and such that their union is F.

7.8 Lemma. If $\equiv$ is a congruence relation on F, then there exists a partition of F into sets (called congruence classes) such that the elements in each congruence class are in the relation $\equiv$ and such that no two elements in different congruence classes are in the relation $\equiv$.

Proof. The congruence classes are simply the sets $\{y: y \equiv x\}$ where x varies over F. Let us denote this set by $x/\equiv$. The reader easily verifies that $y \in x/\equiv$ if and only if $x \equiv y$: Also, it is obvious that $x/\equiv \cap y/\equiv \neq \varnothing$ iff $x/\equiv \;=\; y/\equiv$ and iff $x \equiv y$. From this it follows that the set of all $x/\equiv$ forms a partition of F. Let us, by abuse of notation, denote the set of congruence classes by $F/\equiv$.

Remark. We did not actually need property (iv), thus the lemma holds also for relations satisfying (i)–(iii) above. Such relations are called *equivalence relations*.

Let us now define a new finite computer as follows. The set of states is the set $F/\equiv$ of all congruence classes of F with respect to the congruence relation $\equiv$ defined above. The start state t_0 of $F/\equiv$ is $s_0/\equiv$. The transition functions g_{a_i} are defined by $g_{a_i}(s/\equiv) = f_{a_i}(s)/\equiv$. By property (iv) of the relation $\equiv$ this definition of g_{a_i} is consistent. Namely, if $s' \equiv s$, then

$$g_{a_i}(s/\equiv) = f_{a_i}(s)/\equiv\; =\; f_{a_i}(s')/\equiv\; =\; g_{a_i}(s'/\equiv),$$

which shows that the definition of g_{a_i} does not depend on the particular element s chosen in the congruence class $s/\equiv$.

Consider now the finite automaton $\langle G, t_0, g_a, \ldots, g_{a_m} \rangle$. We may assume that each state t in G is *accessible*. By this we mean that there is for every $t \in G$ a word $w \in A^*$ such that $G(w) = t$. In the case the state t happens not to be accessible we eliminate it from G. The result of eliminating all nonaccessible states from G is called the *connected part* of the automaton and denoted by $(\langle G, t_0, g_a, \ldots, g_{a_m} \rangle, \Psi)^c$. Obviously $(\langle G, t_0, g_{a_1}, \ldots, g_{a_m} \rangle, \Psi)^c$ and $(\langle G, t_0, g_a, \ldots, g_{a_m} \rangle, \Psi)$ are equivalent.

We still have not described what the output assignment of $\langle G, t_0, g_{a_1}, \ldots, g_{a_m} \rangle$ is to be. Let us define $\Psi: G \to B$ by $\Psi(s/\equiv) = \Phi(s)$. This definition again does not depend on the choice of s. Namely, suppose that $s_1, s_2 \in s/\equiv$. Then $s_1 \equiv s_2$, and hence $\Phi_{s_1}(w) = \Phi_{s_2}(w)$ for all $w \in A^*$. So in particular $\Phi_{s_1}(\lambda) = \Phi(s_1) = \Phi(s_2) = \Phi_{s_2}(\lambda)$.

Thus the computer

$$(\langle G, t_0, g_a, \ldots, g_{a_m} \rangle, \Psi) = (\langle F, s_0, f_{a_1}, \ldots, f_{a_m} \rangle, \Phi)/\equiv$$

is well defined. The main reason for describing the construction above is that the constructed computer is equivalent to the given one; indeed, it is the sought-for minimal computer. The proper notion to introduce in order to get at these facts is that of a homomorphism between computers.

7.9 Definition. Let $(\langle F, s_0, f_{a_1}, \ldots, f_{a_m} \rangle, \Phi)$ and $(\langle G, t_0, g_{a_1}, \ldots, g_{a_m} \rangle, \Psi)$ be finite computers with the same input and output alphabets. Let h be a map of F onto G. Then h is called a *homomorphism* of the first computer onto the second if it satisfies the following requirements:

(1) $h(s_0) = t_0$;
(2) $h(f_{a_i}(s)) = g_{a_i}(h(s))$ for all $s \in F$, $a_i \in A$;
(3) $\Psi(h(s)) = \Phi(s)$ for all $s \in F$.

Example. The map j which associates to each state s its equivalence class is a homomorphism. Namely

(1) $j(s_0) = s_0/\equiv\, = t_0$,
(2) $j(f_{a_i}(s)) = f_{a_i}(s)/\equiv\, = g_{a_i}(s/\equiv) = g_{a_i}(j(s))$,
(3) $\Psi(j(s)) = \Psi(s/\equiv) = \Phi(s)$.

7.10 Theorem. If h is a homomorphism of the finite computer $(\langle F, s_0, f_{a_1}, \ldots, f_{a_m}\rangle, \Phi)$ onto $(\langle G, t_0, g_{a_1}, \ldots, g_{a_m}\rangle, \Psi)$, then the two computers are equivalent.

Proof. We first establish that $G(w) = h(F(w))$ for all w by induction on w. Observe that $G(\lambda) = t_0 = h(s_0) = h(F(\lambda))$ by definition of F, G, and property (1) of homomorphisms. Moreover,

$$G(wa_i) = g_{a_i}(G(w)) = g_{a_i}(h(F(w)))$$

by induction assumption; $g_{a_i}(h(F(w))) = h(f_{a_i}(F(w)))$ by property (2) of homomorphisms; $h(f_{a_i}(F(w))) = h(F(wa_i))$ by definition of F. So altogether $G(w_{a_i}) = h(F(wa_i))$.

The identity $G(w) = h(F(w))$ allows us to prove the identity $\Phi(w) = \Psi(w)$ as follows, by induction on w. First observe that

$$\Psi(\lambda) = \Psi(t_0) = \Psi(h(s_0)) = \Phi(s_0) = \Phi(\lambda)$$

by definition of Φ, Ψ and property (3) of homomorphisms. The induction step is accomplished by observing

$$\Psi(wa_i) = \Psi(w) \cdot \Psi(G(wa_i)) = \Phi(w) \cdot \Psi(hF(wa_i))$$

$$= \Phi(w) \cdot \Phi(F(wa_i)) = \Phi(wa_i).$$

Corollary. The computers $(\langle F, s_0, f_{a_1}, \ldots, f_{a_m}\rangle, \Phi)$ and $(\langle F, s_0, f_{a_1}, \ldots, f_{a_i}\rangle, \Phi)/\equiv$ are equivalent.

7.11 Definition. A computer $(\langle G, t_0, g_{a_1}, \ldots, g_{a_m}\rangle, \Psi)$ is called *minimal* if for every computer $(\langle F, s_0, f_{a_1}, \ldots, f_a\rangle, \Phi_m)$ which is equivalent to it there exists a homomorphism of its connected part $(\langle F, s_0, f_{a_1}, \ldots, f_{a_m}\rangle, \Phi)^c$ onto $(\langle G, t_0, g_{a_1}, \ldots, g_{a_m}\rangle, \Psi)$.

7. FINITE TRANSDUCERS, MINIMALIZATION

7.12 Theorem. The computer

$$(\langle G, t_0, g_{a_1}, \ldots, g_{a_m}\rangle, \Psi) = (\langle F, s_0, f_{a_1}, \ldots, f_{a_m}\rangle, \Phi)^c/\equiv$$

is minimal.

Proof. Suppose that $(\langle H, r_0, h_{a_1}, \ldots, h_{a_m}\rangle, \Gamma)$ is equivalent to $(\langle F, s_0, f_{a_1}, \ldots, f_{a_m}\rangle, \Phi)$. Let r be any state in the connected part of H. By equivalence, there is a state s in the connected part of F which is equivalent to r. Let $h(r) = s/\equiv$ for some such s. The function h, thus defined, is a homomorphism. Note first that $h(r_0) = s_0/\equiv\, = t_0$. Also, $h_{a_i}(r)$ is equivalent to $f_{a_i}(s)$ by Lemma 7.6, hence

$$h(h_{a_i}(r)) = f_{a_i}(s)/\equiv\, = g_{a_i}(s/\equiv) = g_{a_i}(h(r)).$$

Condition (3) is easily verified. It follows that $h(H(w)) = G(w)$ for all $w \in A^*$. Namely, $h(H(\lambda)) = h(r_0) = s_0/\equiv\, = t_0 = G(\lambda)$, since r_0 and s_0 are equivalent.

$$h(H(w \cdot a_i)) = h(h_{a_i}(H(w))) = g_{a_i}(h(H(w))) = g_{a_i}(G(w)) = G(w \cdot a_i),$$

since h is a homomorphism. Now observe that the connected part of $(\langle H, r_0, h_{a_1}, \ldots, h_{a_m}\rangle, \Gamma)$ consists of those states which are of the form $H(w)$; this set of states is mapped by h onto the set of states $G(w)$, $w \in A^*$, that is, on the set of all states of $(\langle G, t_0, g_{a_1}, \ldots, g_{a_m}\rangle, \Psi)$, by construction of this computer. Thus h has the right properties and minimality is established.

7.13 Theorem. For each finite computer, there exists a minimal computer which is equivalent to it. This minimal computer is unique (up to isomorphism).

Proof. The existence of a minimal computer has been established in the previous theorem. There remains to show uniqueness; that is, we have to construct an *isomorphism* between any two equivalent minimal computers. We define a homomorphism between two computers as an isomorphism if it is one-to-one and onto as a map between the states. Suppose then that $(\langle G_1, t_0^1, g_a^1, \ldots\rangle, \Psi^1)$ and $(\langle G_2, t_0^2, g_{a_1}^2, \ldots\rangle, \Psi^2)$ are two equivalent computers. By assumption there are maps $h_1: G_1 \to G_2$ (onto) and $h_2: G_2 \to G_1$ (onto) which are both homomorphisms. Since both G_1 and G_2 are finite, it follows that both h_1 and h_2 are one-to-one.

7.14 Problems
(a) Show that $\Phi_s(u \cdot v) = \Phi_s(u) \cdot \Phi_{F_s(u)}(v)$.
(b) Suppose that h is a homomorphism between two computers. Define $s_1 \approx s_2$ by $h(s_1) = h(s_2)$. Show that $\approx$ is a congruence relation on the first computer.

Let $(\langle F, s_0, f_{a_1}, \ldots, f_{a_m}\rangle, \Phi)$ be a given finite computer, and suppose we are asked to construct the minimal computer equivalent to it. The proof of the existence of such a minimal computer (Theorem 7.13) is apparently nonconstructive in two places:

(a) In determining whether two states s_1 and s_2 of F are in the relation $\equiv$, we have to make sure that $\Phi_{s_1}(w) = \Phi_{s_2}(w)$ for all words $w \in A^*$, that is, for infinitely many words. Checking one word after the other would certainly not be the right way to go about this.

(b) In determining whether a state t of G is inaccessible, we have to prove that $G(w) \neq t$ for all $w \in A^*$. Again, this is a process that is apparently not effective.

The fact is, however, that both these questions can be solved by finite, well-determined processes.

7.15 Definition. Consider a finite computer F with output assignment Φ. Two states, s_1 and s_2, of F are called k-equivalent, in symbols $s_1 \equiv_k s_2$, if $\Phi_{s_1}(w) = \Phi_{s_2}(w)$ for all words w of length k.

The relations $\equiv_k$ are equivalence relations, but not necessarily congruence relations. (Why?) If $k' > k$, then $\equiv_{k'}$ may subdivide a $(\equiv_k)$-equivalence class of states in F. We say that $\equiv_{k'}$ is a refinement of $\equiv_k$. Since F is finite we can only have a finite number of proper refinements; that is, a point will be reached when $\equiv_k$ equals $\equiv_{k'}$ for all $k' \geqslant k$. Our next lemma tells us a useful criterion for finding such k.

7.16 Lemma. If $\equiv_k$ coincides with $\equiv_{k+1}$, then $\equiv_k$ is the same as $\equiv_n$ for all $n \geqslant k$; hence $\equiv_k$ equals $\equiv$ in that case.

7. FINITE TRANSDUCERS, MINIMALIZATION

Proof. Let us assume that $\equiv_{k+1}$ equals $\equiv_k$. We want to show that $\equiv_k$ equals $\equiv_n$ for all $n \geq k$ by induction. So suppose that $\equiv_k$ equals $\equiv_{k'}$ for all $k' \leq n$, in particular $\equiv_{n-1}$ equals $\equiv_n$, and that $s \equiv_n t$. We have to show $s \equiv_{n+1} t$. For this we first prove $f_{a_i}(s) \equiv_n f_{a_i}(t)$ for each $a_i \in A$. Namely, let w be a word of length $n - 1$. Then $\Phi_{f_{a_i}(s)}(w) = d(\Phi_s(a_i w))$ by the observation in the proof of Lemma 7.6. Since $s \equiv_n t$ by assumption, we have $d(\Phi_s(a_i w)) = d(\Phi_t(a_i w))$, which is $\Phi_{f_{a_i}(t)}(w)$. Thus $f_{a_i}(s) \equiv_{n-1} f_{a_i}(t)$ by definition of $\equiv_{n-1}$. Since $\equiv_{n-1}$ equals $\equiv_n$ by assumption, we conclude that $f_{a_i}(s) \equiv_n f_{a_i}(t)$. We let now w be a word of length n and observe the following:

$$\Phi_s(a_i w) = \Phi_s(a_i) \cdot d(\Phi_{f_{a_i}(s)}(w))$$
$$= \Phi_t(a_i) \cdot d(\Phi_{f_{a_i}(t)}(w)) = \Phi_t(a_i w) \quad \text{for all} \quad a_i.$$

By definition, it follows that $s \equiv_{n+1} t$.

7.17 Lemma. If a finite computer F has n states, then two states, s_1 and s_2, of F are equivalent if and only if $\Phi_{s_1}(w) = \Phi_{s_2}(w)$ for all words w of length $n - 1$.

Proof. We show that $\equiv_{n-1}$ equals $\equiv$. By Lemma 7.16, it is sufficient to show that $\equiv_k$ equals $\equiv_{k+1}$ for some $k \leq n - 1$. For this purpose, note that if $\equiv_{k+1}$ differs from $\equiv_k$, then $\equiv_{k+1}$ has at least one more equivalence class than $\equiv_k$. Now, the number of equivalence classes is bounded by n, the number of elements out of which we form the equivalence classes. If $\equiv_0$ has only one equivalence class, then obviously $\equiv_0$ equals $\equiv$ and we are done. If $\equiv_0$ has at least two equivalence classes and if $\equiv_{n-1}$ differs from $\equiv_{n-2}$, then $\equiv_{n-1}$ will have at least n equivalence classes (since $\equiv_t$ will have at least $k + 1$ equivalence classes for $k = 0, 1, \ldots, n - 1$).

Lemma 7.17 gives us an effective procedure for determining the equivalence of states, since all we need to do is check the finitely many words of length $n - 1$ for $\Phi_{s_1}(w) = \Phi_{s_2}(w)$. It remains to show how the accessible states can be singled out.

7.18 Lemma. If a finite computer F has n states and the state s is accessible, then $s = F(w)$ for some word of length at most $n - 1$.

Proof. Suppose that the shortest word w for which $F(w) = s$ has length at least n. Consider the set of states attained successively by the computer while processing the word w. One of the states must be attained at least twice in this process; that is, there must be a nonempty word y such that $w = x \cdot y \cdot z$ and $F(x) = F(x \cdot y)$. But then $F(x \cdot z) = s$ and length $(x \cdot z)$ is smaller than length (w), a contradiction.

Thus, again, there remain only finitely many words to check, and we have proved the lemma.

7.19 Theorem. There is an effective procedure for minimalizing finite computers. Moreover, this procedure can be used to decide whether two given finite computers are equivalent.

Namely, two computers are equivalent if and only if their corresponding minimal computers are isomorphic. The process of obtaining a minimal computer is effective, and given two computers it is easy to decide whether they are isomorphic.

7.20 Example. Show that the computers F_1 (Fig. 1.29) and F_2 (Fig. 1.30) are equivalent.

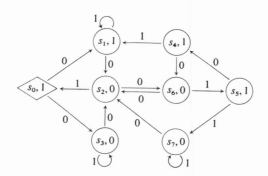

Fig. 1.29. Computer F_1.

Solution. We first find the minimal computer for F_1. For this, we

7. FINITE TRANSDUCERS, MINIMALIZATION

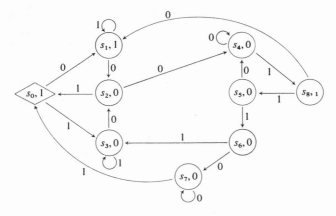

Fig. 1.30. *Computer F_2.*

establish the tabulation given in Table 1.5. Observe that the following are the equivalence classes for $\equiv_i$, $i = 0, 1, 2$.

$\equiv_0$: $\{s_0, s_1, s_4, s_5\}$, $\quad \{s_2, s_3, s_6, s_7\}$;

$\equiv_1$: $\{s_0, s_5\}$, $\quad \{s_1, s_4\}$, $\quad \{s_2, s_6\}$, $\quad \{s_3, s_7\}$;

$\equiv_2$: $t_0 = \{s_0, s_5\}$, $\quad t_1 = \{s_1, s_4\}$, $\quad t_2 = \{s_2, s_6\}$, $\quad t_3 = \{s_3, s_7\}$.

Thus the two relations $\equiv_1$ and $\equiv_2$ are equal. If this is the case, then $\equiv_1$ is the same as the relation $\equiv$ by Lemma 7.16. It follows that the minimal computer for F_1 is that shown in Fig. 1.31.

TABLE 1.5

w	$\Phi_{s_0}(w)$	$\Phi_{s_1}(w)$	$\Phi_{s_2}(w)$	$\Phi_{s_3}(w)$	$\Phi_{s_4}(w)$	$\Phi_{s_5}(w)$	$\Phi_{s_6}(w)$	$\Phi_{s_7}(w)$
λ	1	1	0	0	1	1	0	0
0	11	10	00	00	10	11	00	00
1	10	11	01	00	11	10	01	00
00	110	100	000	000	100	110	000	000
01	111	101	001	001	101	111	001	001
10	100	110	011	000	110	100	011	000
11	100	111	010	000	111	100	010	000

TABLE 1.6

w	$\Phi_{s_0}(w)$	$\Phi_{s_1}(w)$	$\Phi_{s_2}(w)$	$\Phi_{s_3}(w)$	$\Phi_{s_4}(w)$	$\Phi_{s_5}(w)$	$\Phi_{s_6}(w)$	$\Phi_{s_7}(w)$	$\Phi_{s_8}(w)$
λ	1	1	0	0	0	0	0	0	1
0	11	10	00	00	00	00	00	00	11
1	10	11	01	00	01	00	00	01	10
00	110	100	000	000	000	000	000	000	110
01	111	101	001	001	001	001	001	001	111
10	100	110	011	000	011	000	000	011	100
11	100	111	010	000	010	000	000	010	100

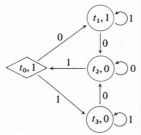

Fig. 1.31. Minimal computer for F_1.

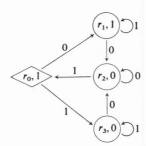

Fig. 1.32. Minimal computer for F_2.

Let us now find the minimal computer for F_2 by the same process. First, we set up the tabulation given in Table 1.6. The equivalence classes are therefore as follows.

$\equiv_0$: $\{s_0, s_1, s_8\}$, $\{s_2, s_3, s_4, s_5, s_6, s_7\}$;

$\equiv_1$: $\{s_0, s_8\}$, $\{s_1\}$, $\{s_2, s_4, s_7\}$, $\{s_3, s_5, s_6\}$;

$\equiv_2$: $r_0 = \{s_0, s_8\}$, $r_1 = \{s_1\}$, $r_2 = \{s_2, s_4, s_7\}$, $r_3 = \{s_3, s_5, s_6\}$.

We obtain the minimal computer shown in Fig. 1.32. Since the two minimal computers for F_1 and F_2 are obviously isomorphic, the two original computers, F_1 and F_2 are equivalent.

7.21 Problems
(a) Two finite *automata* over the same alphabet are called *equivalent* if they accept the same sets of words. Modify the approach in this section to deal with automata and show that the question as to whether two automata are equivalent is decidable. (Hence, so is the problem of the equivalence of regular grammars.)

(b) Think about the number of steps necessary to minimize a given computer. [A typical step is to compute $f_{a_i}(s)$ for given a_i and s.] The process described above is somewhat wasteful. It can easily be streamlined so as to obtain a number $f(n)$ of steps that is proportional to n^2 where n is the number of states of the given computer. Hopcroft [13] has produced an algorithm for which $f(n)$ is proportional to $n \cdot \log n$. Describe the best algorithm you can find and estimate $f(n)$ for it.

8.* TRUTH-FUNCTIONS

This section is a rather naive account on how one could actually *realize* an abstractly given finite computer by wiring up very simple electronic devices. The naïveté lies in the fact that actual computers are much larger and much more complicated than the finite transducers that we are considering here. Also, we deal with some rather special electronic devices here, idealized ones at that, and disregard entirely the very real technological problems connected with realizations.

But in order to be able to handle the more complicated situations of "real life," it is advantageous to learn first to understand the simplest cases thoroughly. Moreover, the mathematics that we need for these simple cases is quite interesting in itself.

Imagine, then, the problem of constructing a device into which we can feed, symbol by symbol, a given word over some finite alphabet A and which will emit such symbols at each turn. To be concrete, let us consider the finite transducer shown in Fig. 1.33. The goal is to provide details in a picture like the one given in Fig. 1.34. Let us assume that

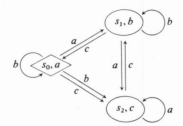

Fig. 1.33

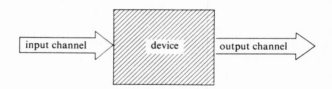

Fig. 1.34

input can only be given in terms of a sequence of pulses along one or more input wires (constituting the input channel). For example, if we have two input wires x_1 and x_2, we may have the situation depicted in Fig. 1.35. Now, two input wires with pulses as illustrated can easily serve to mimic the input of symbols from an alphabet of up to four symbols.

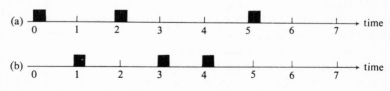

Fig. 1.35. Input wires (a) x_1 and (b) x_2.

8. TRUTH-FUNCTIONS

Let us represent a pulse by writing 1, the absence of a pulse by writing 0. Then each symbol from our input alphabet $A = \{a, b, c\}$ is encoded by a pair of values 0 or 1 assigned to x_1 and x_2. To be concrete, let us take

	x_1	x_2
a	0	0
b	0	1
c	1	0

The sequence of pulses illustrated in Fig. 1.35 thus corresponds to an input word *cbcbbcaa*, for example.

The output will be organized in a similar manner, in our case we will need two output wires, say y_1 and y_2, and we shall use the same encoding as the one above. We have thus arrived at a device that looks like the one shown in Fig. 1.36 where we now know that the device will act on pulses and nonpulses on the two input wires and emit the same kind of signals through the output wires.

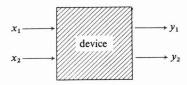

Fig. 1.36

Remember that the device is capable only of finitely many states. These states will be technologically realized by states of the device. A state of the device, however, is nothing else but the description of the presence or absence of pulses at various internal locations. If there are only three states, as in our examples, it is enough to describe the absence or presence of pulses at two locations l_1 and l_2. For example,

State	l_1	l_2
s_0	0	1
s_1	1	0
s_2	1	1

The device should now be constructed to do two things. First, given any state, it should produce its corresponding output. Second, given any state and input it should switch to the next state according to our state diagram. Let us consider the first part: The state is encoded by the presence or absence of a pulse at locations l_1 and l_2, the output by the presence or absence of a pulse at the output wires y_1 and y_2. So there must be two parts of the output device, one that produces the output at y_1 and the other at y_2, both being connected to locations l_1 and l_2. Our device already has more detail, as shown in Fig. 1.37.

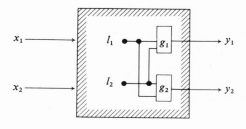

Fig. 1.37

Next we have to realize the next-state function. At time t we obtain a pulse or nonpulse at x_1 and x_2 (representing the incoming symbol at time t); also at time t we have a pulse or nonpulse at locations l_1 and l_2 (representing the state at time t). We would like to have a pulse or nonpulse at locations l_1 and l_2 at time $t + 1$ [representing the state at the time (instant $t + 1$) when the next symbol is read in]. So again, there must be two next-state devices, one that produces an output at l_1 the other at l_2,

Fig. 1.38. *Delay line.*

and both of these outputs should be delayed by one time unit. Let us imagine a device, called delay line, which brings forth at time $t + 1$ exactly what entered it at time t. We will represent it graphically as shown in Fig. 1.38. Apart from this delay line, each next-state device f_1 and f_2 will have four input wires: two that connect up to x_1 and x_2 and two that connect to l_1 and l_2. Altogether our device now looks like the diagram presented in Fig. 1.39.

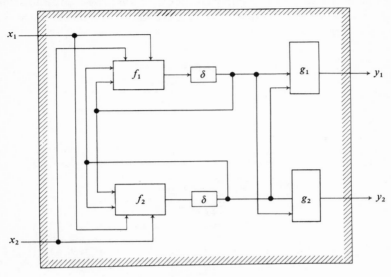

Fig. 1.39

Now each of the internal devices g_1, g_2, f_1, f_2 is to have a rather simple behavior: Depending on pulses or nonpulses on their input wires, they produce a pulse at their respective output wires at exactly the same instant. Mathematically, these devices realize functions (of two, respectively, four arguments) from $\{0,1\}$ to $\{0,1\}$. Such functions are called *truth-functions*, and their study is the main tool in finishing up our task of realizing the given finite transducer. Tables 1.7 and 1.8 contain the data for the functions as determined from the state diagram. Note that these tables are incomplete (in Table 1.7 the case $l_1 = l_2 = 0$ is missing since there is no state corresponding to it). The behavior of the g_i, f_j could be arbitrarily chosen for such argument, without changing the behavior of the computer that we are constructing. Technological considerations may determine how we want to complete the tables. We now turn our backs on technology and begin to treat the mathematics that has become involved here.

Let us fix, once and for all, a set $\{0,1\}$ of two elements, called truth-values: 0 will often be interpreted as "false" and 1 as "true." But at the present time this is of no importance. By a *truth-function* we understand a function (of any number of variables) whose arguments range over truth-values and whose values are truth-values.

TABLE 1.7
Output Functions

l_1	l_2	$g_1(l_1, l_2)$	$g_2(l_1, l_2)$
0	1	0	0
1	0	0	1
1	1	1	0

TABLE 1.8
Next-State Functions

x_1	x_2	l_1	l_2	$f_1(x_1,x_2,l_1,l_2)$	$f_2(x_1,x_2,l_1,l_2)$
0	0	0	1	1	0
0	1	0	1	0	1
1	0	0	1	1	1
0	0	1	0	1	1
0	1	1	0	1	0
1	0	1	0	0	1
0	0	1	1	1	1
0	1	1	1	0	1
1	0	1	1	1	0

8.1 Definition.[9] Let n be a positive integer. An n-ary truth-function f is a function

$$f: \underbrace{\{0,1\} \times \cdots \times \{0,1\}}_{n \text{ times}} \to \{0,1\}.$$

[9] The definitive theory of truth-functions is in Post (18).

8. TRUTH-FUNCTIONS

Since truth-functions have finite domains, it is feasible to present them in the form of tables. Some of them are important enough to have a standard symbol assigned to them.

Negation: ¬

x	$\neg(x)$
0	1
1	0

Conjunction: ∧

x	y	$x \wedge y$
0	0	0
0	1	0
1	0	0
1	1	1

(Note that we write the function symbol between the two argument symbols as customary for functions of two variables—to wit, +, ·.)

Disjunction: ∨

x	y	$x \vee y$
0	0	0
0	1	1
1	0	1
1	1	1

Conditional: ⊃

x	y	$x \supset y$
0	0	1
0	1	1
1	0	0
1	1	1

Biconditional: ≡

x	y	x ≡ y
0	0	1
0	1	0
1	0	0
1	1	1

Sheffer's Function: |

x	y	x|y
0	0	1
0	1	1
1	0	1
1	1	0

These functions can be combined in the usual manner to produce new functions, for example, $f(x,y,z) = \neg((x \wedge y) \supset z)$. We shall do this by giving a position of central importance to some rather trivial functions namely the projection functions, and to the composition of truth-functions.

This fact becomes rather important if we are asked to realize complicated truth-functions (such as g_1, g_2, f_1, f_2 in our original example) by means of simple devices.[10] For example, to realize the function $f(x,y,z) = \neg((x \wedge y) \supset z)$ we can make use of devices which realize $\neg, \wedge,$ and $\supset$, respectively, as depicted in Fig. 1.40.

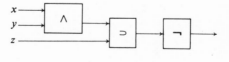

Fig. 1.40

[10] Questions of this nature are treated, for example, in Korfhage (15). The interested reader may wish to follow this up and learn the elementary parts of Boolean algebra which was first investigated by Boole (10) in connection with logic.

8. TRUTH-FUNCTIONS

8.2 Definition

(a) The *i*th *n*-ary projection function U_i^n is the function

$$U_i^n: \underbrace{\{0,1\} \times \cdots \times \{0,1\}}_{n \text{ times}} \to \{0,1\}$$

defined by $U_i^n(x_1,\ldots,x_n) = x_i$. (Note that $1 \leqslant i \leqslant n$.)

(b) Suppose that f is an *n*-ary truth-function and $g_1, \ldots, g_n$ are *m*-ary truth-functions. Then $f(g_1,\ldots,g_n)$ is the *m*-ary truth-function defined by

$$f(g_1,\ldots,g_n)(x_1,\ldots,x_m) = f(g_1(x_1,\ldots,x_m),\ldots,g_n(x_1,\ldots,x_m)).$$

The function $f(g_1,\ldots,g_n)$ is called the composition of $g_1, \ldots, g_n$ by means of f.

For example, the function customarily defined by

$$f(x,y,z) = ((x \wedge y) \vee ((x \wedge z) \supset (\neg x)))$$

is obtained from $\wedge, \vee, \supset, \neg$, and the U_i^n as follows by composition only:

$$f = \vee(\wedge(U_1^3, U_2^3), \supset(\wedge(U_1^3, U_3^3), \neg(U_1^3)))$$

or, if we write the binary function symbols between the arguments,

$$f = (U_1^3 \wedge U_2^3) \vee ((U_1^3 \wedge U_3^3) \supset \neg(U_1^3)).$$

The advantage of the latter formulation over the former is that we have eliminated the mention of variables, and the places where they occur in a formula, from the definition of the composite function. This serves to better bring out the composite nature of the defined function. It also allows us to avoid mentioning variables at all when defining the totality of functions that can be obtained by combining functions from a given set such as $\{\wedge, \neg, \supset\}$. Let us now give a formal definition of this concept.

8.3 Definition. Let $\mathscr{T}$ be the set of all truth-functions, $\mathscr{T}_n$ the set of all *n*-ary truth-functions, $n = 1, 2, \ldots$, and let G be any set of truth-functions. The set of truth-functions generated by G, in symbols $[G]$, is obtained as follows:

(i) Each function f in G is also in $[G]$.
(ii) All projection functions p_n^i belong to $[G]$.
(iii) If $f \in \mathcal{T}_n$ belongs to $[G]$ and $g_1, \ldots, g_n \in \mathcal{T}_m$ belong to $[G]$, then $f(g_1,\ldots,g_n)$ belongs to $[G]$.
(iv) No other truth-functions belong to $[G]$.

We should now like to obtain some more insight into the structure of the set $[G]$ generated by G. The goal is to answer such questions as to whether a given truth-function f belongs to $[G]$ when we only know what G is.

8.4 Definition. A sequence of sets of truth-functions $G_0, G_1, G_2, \ldots$, is called a *generation sequence* with *generators* G if G_0 is the set of all projection functions U_i^n, $n = 1, 2, \ldots$; G_{k+1} is the union of G_k and the set of all functions $f(g_1,\ldots,g_n)$ where $f \in G \cap \mathcal{T}_n$ and $g_1, \ldots, g_n \in G_k \cap \mathcal{T}_m$ for some n and m.

Observe that $G_0 \subseteq G_1 \subseteq G_2 \subseteq \cdots$. Let us now prove the main fact about generation sequences.

8.5 Theorem. If $G_0, G_1, G_2, \ldots$, is a generation sequence with generators G, then $[G] = G_0 \cup G_1 \cup G_2 \cup \cdots$.

Proof
(a) $[G] \subseteq G_0 \cup G_1 \cup G_2 \cup \cdots$. To prove this fact, we need to show that $G_0 \cup G_1 \cup G_2 \cup \cdots$ has properties (i)–(iii) of Definition 8.3. Property (i) follows from the fact that $G \subseteq G_1$: If $f \in G \cap \mathcal{T}_n$, then

$$f = f(U_1^n, U_2^n, \ldots, U_n^n) \in G_1.$$

Property (ii) is obvious from the definition of generation sequences. To verify property (iii), we first prove a weaker assertion: If $f \in G \cap \mathcal{T}_n$ and $g_1,\ldots,g = \mathcal{T}_m \cap (G_0 \cup G_1 \cup \cdots)$, then $f(g_1,\ldots,g_n) \in G_0 \cup G_1 \cup G_2 \cup \cdots$. Namely, in this case $g_1 \in G_{i_1}$, $g_2 \in G_{i_2}$, $\ldots$, $g_n \in G_{i_n}$ for some natural numbers $i_1, \ldots, i_n$. Let k be the maximum of $i_1, \ldots, i_n$. Then $g_1, \ldots, g_n \in G_k$ and $f(g_1,\ldots,g_n) \in G_{k+1}$ by the definition of generating sequences. Hence $f(g_1,\ldots,g_n) \in G_0 \cup G_1 \cup G_2 \cup \cdots$. This proves a weakened form of property (iii). For the full content of (iii), we have to assume $f \in G_r \cap \mathcal{T}_n$

8. TRUTH-FUNCTIONS

instead of $f \in G \cap \mathcal{T}_n$. Let us proceed by induction on r. If $r = 0$ and $f \in G_0 \cap \mathcal{T}$, then $f = U_i^n$ for some i. In this case, we have

$$f(g_1, \cdots, g_n) = g_i \in G_0 \cup G_1 \cup \cdots.$$

Suppose that (iii) is true for r; that is, suppose that whenever $h \in G_r \cap \mathcal{T}_s$ and $g_1, \cdots, g_s \in \mathcal{T}_m \cap (G_0 \cup G_1 \cup \cdots)$, then $h(g_1, \cdots, g_s) \in (G_0 \cup G_1 \cup \cdots)$. Let $h \in G_{r+1} \cap \mathcal{T}_m$. If $h \in G_r$, then there is nothing to prove. Otherwise, h is of the form $h = f(h_1, \ldots, h_n)$ with $f \in G \cap \mathcal{T}_n$ and $h_1, \cdots, h_n \in G_r \cap \mathcal{T}_s$. The function that we are considering is $h(g_1, \ldots, g_s)$; it takes the form $(f(h_1, \ldots, h_n))(g_1, \ldots, g_s)$. We now make the important observation that

$$(f(h_1, \ldots, h_n))(g_1, \ldots, g_s) = f(h_1(g_1, \ldots, g_s), \ldots, h_n(g_1, \ldots, g_s)).$$

This equation follows from Definition 8.2. Namely, for any

$$x_1, \cdots, x_m \in \{0, 1\},$$

the left-hand side is

$$(f(h_1, \ldots, h_n)(g_1, \ldots, g_s)(x_1, \ldots, x_m))$$
$$= f(h_1, \ldots, h_n)(g_1(x_1, \ldots, x_m), \ldots, g_s(x_1, \ldots, x_m))$$
$$= f(h_1(g_1(x_1, \ldots, x_m), \ldots, g_s(x_1, \ldots, x_m)), \ldots, h_n(g_1(x_1, \ldots, x_s), \ldots, g_s(x_1, \ldots, x_m))).$$

The right-hand side is

$$f(h_1(g_1, \ldots, g_s), \ldots, h_n(g_1 \ldots, g_s))(x_1, \ldots, x_m)$$
$$= f(h_1(g_1, \ldots, g_s)(x_1, \ldots, x_m), \ldots, h_n(g_1, \ldots, g_s)(x_1, \ldots, x_m))$$
$$= f(h_1(g_1(x_1, \ldots, x_m), \ldots, g_s(x_1, \ldots, x_m)), \ldots, h_n(g_1(x_1, \ldots, x_m), \ldots, g_s(x_1, \ldots, x_m))).$$

Thus the two sides are equal, and we have

$$h(g_1, \ldots, g_s) = f(h_1, \ldots, h_n)(g_1, \ldots, g_s)$$
$$= f(h_1(g_1, \ldots, g_s), \ldots, h_n(g_1, \ldots, g_s)).$$

From this last form of the function in question we conclude at once, using the induction hypothesis and the weakened version of (iii), that $h(g_1,\cdots,g_s) \in G_0 \cup G_1 \cup \cdots$. It follows that $G_0 \cup G_1 \cup G_2 \cup \cdots$ has properties (i)–(iii), and hence $[G] \subseteq G_0 \cup G_1 \cup G_2 \cup \cdots$.

(b) $G_0 \cup G_1 \cup G_2 \cup \cdots \subseteq [G]$. This fact is easily proved by showing $G_k \subseteq [G]$ using induction on k. Namely, $G_0 \subseteq [G]$ by (ii). If $G_k \subseteq [G]$, then $G_{k+1} \in [G]$. For suppose that $h \in G_{k+1}$. If $h \in G_k$, then $h \in [G]$ by induction assumption. If $h = f(g_1,\cdots,g_n) \in G_{k+1}$, where $f \in G \cap \mathscr{T}_n$ and $g_1, \cdots, g_n \in G_k \cap \mathscr{T}_m$, then $g_1, \cdots, g_n \in [G]$ and hence $f(g_1,\cdots,g_n) \in [G]$ by (iii). Hence $G_k \subseteq [G]$ for all k and therefore $G_0 \cup G_1 \cup G_2 \cup \cdots \subseteq [G]$.

8.6 Theorem. If $G_0, G_1, G_2, \ldots$, is a generation sequence with generators G, then for any m we have

(a) $G_{k+1} \cap \mathscr{T}_m$ is the union of $G_k \cap \mathscr{T}_m$ and the set of all functions $f(g_1,\ldots,g_n)$ where $f \in G \cap \mathscr{T}_n$ and $g_1, \cdots, g_n \in G_k \cap \mathscr{T}_m$.
(b) If $G_k \cap \mathscr{T}_m = G_{k+1} \cap \mathscr{T}_m = H$, then $G_s \cap \mathscr{T}_m = H$ for all $s \geqslant k$.

Proof. Part (a) is obvious from the definition of generation sequence. For part (b) we argue as follows. Suppose that s is the smallest number greater than k for which $G_s \cap \mathscr{T}_m \neq H$. Since $G_s \supseteq G_k$, we must therefore have a function $h \in G_s \cap \mathscr{T}_m$ which is not in $G_k \cap \mathscr{T}_m$. Then

$$h = f(g_1,\ldots,g_n)$$

for some $f \in G \cap \mathscr{T}_n$ and $g_1, \ldots, g_n \in G_{s-1} \cap \mathscr{T}_m$. It follows that $g_1,\ldots,g_n \in G_k \cap \mathscr{T}_m$, since $G_k \cap \mathscr{T}_m = G_{s-1} \cap \mathscr{T}_m$. Hence

$$f(g_1,\ldots,g_n) \in G_{k+1} \cap \mathscr{T}_m = H,$$

which contradicts our assumption that $f(g_1,\ldots,g_n) = h \notin H$. This proves the theorem.

Theorems 8.5 and 8.6 provide us with the tools to solve problems of the following nature. Suppose we have a finite set G of truth-functions. Let f be a given m-ary truth-function. Can f be obtained from functions in G; precisely stated, is $f \in [G]$? Our results apply as follows: $f \in [G]$ iff $f \in [G] \cap \mathscr{T}_m$. But $[G] \cap \mathscr{T}_m = (G_0 \cap \mathscr{T}_m) \cup (G_1 \cap \mathscr{T}_m) \cup \cdots$. Since $\mathscr{T}_m$ is a finite set (it has exactly $2^{(2^m)}$ elements), there is a number $k \leqslant 2^{2m}$ such that $G_k \cap \mathscr{T}_m = G_{k+1} \cap \mathscr{T}_m$. In fact, by part (b) of Theorem 8.6 $G_s \cap \mathscr{T}_m = G_k \cap \mathscr{T}_m$ for all $s \geqslant k$. Hence, if $f \in [G]$ at all, then $f \in G_k \cap \mathscr{T}_m$.

8. TRUTH-FUNCTIONS

Thus it is sufficient to construct the finite sequence of finite sets

$$G_0 \cap \mathcal{T}_m \subseteq G_1 \cap \mathcal{T}_m \subseteq \cdots \subseteq G_k \cap \mathcal{T}_m$$

until no more additional elements are obtained. If f is produced in this process, then it is in $[G]$, otherwise it is not.

8.7 Examples and Problems

(a) Let $G = \{\wedge, \vee\}$. Show that $\supset \notin [G]$.

Solution. Let $x = U_1^2$, $y = U_2^2$.

$\mathcal{T}_2 \cap G_0 = \{x, y\}$

$\mathcal{T}_2 \cap G_1 = \{x \wedge x, x \wedge y, x \vee x, x \vee y, y \wedge x, y \wedge y, y \vee x, y \vee y, x, y\}$

$\quad = \{x, x \wedge y, x, x \vee y, x \wedge y, y, x \vee y, y, x, y\}$

$\quad = \{x \wedge y, x \vee y, x, y\}$

$\mathcal{T}_2 \cap G_2 = \{(x \wedge y) \wedge x, (x \wedge y) \wedge y, (x \wedge y) \vee x, (x \wedge y) \vee y,$

$\quad (x \vee y) \wedge x, (x \vee y) \wedge y, (x \vee y) \vee x, (x \vee y) \vee y,$

$\quad (x \wedge y) \wedge (x \wedge y), (x \wedge y) \wedge (x \vee y), (x \wedge y) \vee (x \wedge y),$

$\quad (x \wedge y) \vee (x \vee y), x \wedge y, x \vee y, x, y\}$

$\quad = \{x \wedge y, x \wedge y, (x \wedge y) \vee x, (x \wedge y) \vee y, x, y, x \vee y,$

$\quad x \vee y, x \wedge y, x \wedge y, x \wedge y, x \wedge y, x \wedge y, x \vee y, x, y\}$

$\quad = \{x \wedge y, x \vee y, x, y\} = \mathcal{T}_2 \cap G_1 = \mathcal{T}_2 \cap [G].$

x	y	$x \wedge y$	$x \vee y$	x	y	$\supset$
0	0	0	0	0	0	1
0	1	0	1	0	1	1
1	0	0	1	1	0	0
1	1	1	1	1	1	1

(b) Show that $\vee \notin [\{\neg, \equiv\}]$.
(c) Show that $\neg \in [\{|\}]$ and $\wedge \in [\{|\}]$.
(d) Show that $\vee \in [\{\neg, \wedge\}]$, $\supset \in [\{\neg, \wedge\}]$, and $\equiv \in [\{..., \wedge\}]$.
'(e) Show that $\wedge \in [\{\neg, \vee\}]$, $\wedge \in [\{\neg, \supset\}]$.

Some sets G have the property that they allow the generation of all truth-functions; we call such sets functionally complete. One example is, of course, $G = \mathscr{T}$. We shall find less trivial ones.

8.8 Definition. A set G of truth-functions is called *functionally complete* if $[G] = \mathscr{T}$.

8.9 Theorem. The set $\{\neg, \wedge, \vee\}$ is functionally complete.

The fact that we have a finite set of truth-functions which is functionally complete allows us to decide whether or not any proposed finite set of truth-functions is complete. How?

Proof. There are just four truth-functions of one variable, the identity function $f_1(x) = x$ for all $x \in \{0,1\}$; the function $f_2(x) = 0$ for all $x \in \{0,1\}$; the function $f_3(x) = 1$ for all $x \in \{0,1\}$; and the function $f_4(0) = 1, f_4(1) = 0$. These functions can be composed of $\neg$, $\wedge$, $\vee$, and the projection functions as follows:

$$f_1 = U_1^1, \quad f_2 = U_1^1 \wedge (\neg U_1^1), \quad f_3 = U_1^1 \vee (\neg U_1^1), \quad f_4 = \neg.$$

This proves that all unary truth-functions belong to the set generated by $\neg$, $\vee$, and $\wedge$, in symbols, $\mathscr{T}_1 \subseteq [\{\neg, \wedge, \vee\}]$. We show by induction on n that $\mathscr{T}_n \subseteq [\{\neg, \wedge, \vee\}]$. Since $\mathscr{T} = \mathscr{T}_1 \cup \mathscr{T}_2 \cup \cdots$, this proves the theorem.

Suppose, then, that $\mathscr{T}_n \subseteq [\{\neg, \wedge, \vee\}]$ and let f be any function in $\mathscr{T}_{n+1}$. We use it to define two functions, f_1 and f_2, in $\mathscr{T}_n$. Namely,

$$f_1(x_1, \ldots, x_n) = f(x_1, \ldots, x_n, 0),$$
$$f_2(x_1, \ldots, x_n) = f(x_1, \ldots, x_n, 1),$$

for all $x_1, \cdots, x_n \in \{0, 1\}$. Let us now consider the function

$$\left(f_2(U_1^{n+1}, \ldots, U_n^{n+1}) \wedge U_{n+1}^{n+1}\right) \vee \left(f_1(U_1^{n+1}, \ldots, U_n^{n+1}) \wedge \neg U_{n+1}^{n+1}\right) = g.$$

This function is a function of $n + 1$ variables; by the induction hypothesis and construction we observe that $g \in [\{\neg, \wedge, \vee\}]$. We are finished if we can show that $f = g$. There are two cases to be distinguished for any $x_1, \cdots, x_n \in \{0, 1\}$:

$$g(x_1, \ldots, x_n, 0) = (f_2(x_1, \ldots, x_n) \wedge 0) \vee (f_2(x_1, \ldots, x_n) \wedge \neg 0)$$
$$= 0 \vee (f_1(x_1, \ldots, x_n) \wedge 1) = f_1(x_1, \ldots, x_n) = f(x_1, \ldots, x_n, 0);$$
$$g(x_1, \ldots, x_n, 1) = (f_2(x_1, \ldots, x_n) \wedge 1) \vee (f_1(x_1, \ldots, x_n) \wedge \neg 1)$$
$$= f_2(x_1, \ldots, x_n) \vee (f_1(x_1, \ldots, x_n) \wedge 0)$$
$$= f_2(x_1, \ldots, x_n) = f(x_1, \ldots, x_n, 1).$$

8.10 Corollary. $\{\neg, \wedge\}$, $\{\neg, \vee\}$, $\{\neg, \supset\}$, and $\{|\}$ are functionally complete.

8.11 Problems
(a) Consider the truth function f given by Table 1.9. Compose $f(x, y, z)$ in terms of $\wedge$, $\vee$, and $\neg$. Simplify!

TABLE 1.9
Truth-Function f

x	y	z	$f(x, y, z)$
0	0	0	1
0	0	1	0
0	1	0	1
0	1	1	0
1	0	0	1
1	0	1	0
1	1	0	1
1	1	1	1

(b) Finish providing the details for the realization of the diagram given in Fig. 1.41 by composing truth-functional components,

to obtain realizations of g_1, g_2, f_1, and f_2.

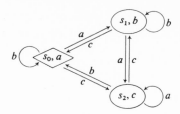

Fig. 1.41

(c) Show that every truth-function $f(x_1,\ldots,x_n)$ can be represented in the form

$$f(x_1,\ldots,x_n) = (a_{1,1} \wedge a_{1,2} \wedge \cdots \wedge a_{1,n}) \vee (a_{2,1} \wedge \cdots \wedge a_{2,n}) \vee \cdots \vee (a_{n,1} \wedge \cdots \wedge a_{n,n})$$

where each $a_{i,j}$ is either of the form x_j or of the form $\neg x_j$. This is called the *disjunctive normal form* for f.

REFERENCES

Books and Collections

[1] M. A. Arbib, *Theories of Abstract Automata*. Prentice-Hall, Englewood Cliffs, New Jersey, 1969.
[2] M. A. Arbib (ed.), *Algebraic Theory of Machines, Languages, and Semigroups*. Academic Press, New York, 1968.
[3] A. Ginzburg, *Algebraic Theory of Automata*. Academic Press, New York, 1968.
[4] M. A. Harrison, *Introduction to Switching and Automata Theory*. McGraw-Hill, New York, 1965.
[5] J. Hartmanis and R. E. Stearns, *Algebraic Structure Theory of Sequential Machines*. Prentice-Hall, Englewood Cliffs, New Jersey, 1966.
[6] M. Minsky, *Computation: Finite and Infinite Machines*. Prentice-Hall, Englewood Cliffs, New Jersey, 1967.
[7] E. F. Moore (ed.), *Sequential Machines: Selected Papers*. Addison-Wesley, Reading, Massachusetts, 1964.
[8] R. J. Nelson, *Introduction to Automata*. Wiley, New York, 1968.
[9] C. E. Shannon and J. McCarthy (eds.), *Automata Studies* (*Ann. of Math. Studies.* No. 34). Princeton Univ. Press, Princeton, New Jersey, 1956.

REFERENCES

Original Articles

[10] G. Boole, *An Investigation of the Laws of Thought* (1854). Dover, New York, 1958.

[11] J. R. Büchi, Mathematische Theorie des Verhaltens endlicher Automaten. *Z. Angew. Math. Mech.*, **42**, 9–16 (1962).

[12] N. Chomsky and G. A. Miller, Finite state languages. *Information and Control*, **1**, 91–112 (1958).

[13] J. Hopcroft, An $n \log n$ algorithm for minimizing states in a finite automaton. Stanford Comp. Sci. Memo No. 190 (1971).

[14] S. C. Kleene, Representation of events in nerve nets and finite automata. In *Automata Studies*, pp. 3–41; see (9).

[15] R. R. Korfhage, *Logic and Algorithms*. Wiley, New York, 1966.

[16] W. S. McCulloch and W. Pitts, A logical calculus of the ideas immanent in nervous activity. *Bull. Math. Biophys.*, **5**, 115–133 (1943).

[17] A. Nerode, Linear automaton transformations. *Proc. Amer. Math. Soc.*, **9**, 541–544 (1958).

[18] E. L. Post, The two-valued iterative systems of mathematical logic. *Ann. Math. Studies*, No. 5. Princeton Univ. Press, Princeton, New Jersey, 1941.

[19] M. O. Rabin and D. Scott, Finite automata and their decision problems. *IBM J. Res. Develop.*, **3**, 115–125 (1959).

[20] R. W. Ritchie, Finite automata and the set of squares. *J. Assoc. Comput. Mach.*, **10**, 528–531 (1963).

[21] A. Salomaa, Axiom systems for regular expressions of finite automata. *Ann. Univ. Turku.*, Ser. A1, **75** (1964).

CHAPTER

Recursive Functions and Programmed Machines

The present chapter presents an introduction into the general theory of computable numerical functions. The main goals are to develop in the reader an understanding of the scope and limitations of this class of functions. The approach is via a model of a computational device, the universal calculator, whose actions are determined by formal programs.[1] In choosing this approach, we were motivated by the wish to prepare, and underline the connection to, the general concept of programmed computation.

The universal calculator and the programs that it accepts have many features that are present in actual scientific computers and their programming languages. These features appear here, of course, in a highly idealized version. We pay some attention to questions of actual realization of machines or execution of programs in the later sections of this chapter. But, in general, our interest is quite theoretical. Thus the class of functions that are termed computable, namely the recursive functions, form a rather extensive class, and for some functions and arguments the computability of the corresponding value may be—while possible in principle—quite out of reach for any conceivable realization.

The present introduction to the theory of recursive functions can, of

---

[1] This approach goes back at least to Hao Wang [16] and Shepherdson and Sturgis [14].

course, not touch upon all important aspects of this theory. In particular, we have not included the interesting connections to mathematical logic. Also, we have treated—apart from the approach through the universal calculator—only one other approach to this theory, namely through Turing machines (and that only briefly). This leaves out the very interesting approaches that are based on the notions of normal algorithm (Markov)[2], λ-conversion (Church), systems of defining equations (Herbrand–Gödel)[3], and the systems of Post.

1. THE UNIVERSAL CALCULATOR

We turn now to devices whose computing power considerably exceeds that of the finite computer. The basic restriction under which finite computers labor is that the "memory" of a finite computer is finite. A finite computer remembers (in the active sense, that its future actions depend on the content of memory) only one thing, namely the state that it is in. A human computer, however, has the advantage of having notepaper handy on which he can commit to memory and later refer to any results of intermediate calculations that he may wish. Of course, it may be argued that the supply of notepaper is finite. Then so is the set of possible texts with which the human computer may cover his supply. Thus it would seem that the human computer after all has only finitely many states available. What is necessary here is a step of idealization, namely to consider the supply of notepaper inexhaustible. To be explicit, let us assume that the computer has available a stock of notepaper sheets labeled by the variables $x_1, x_2, x_3, \ldots$. During computations any natural number may appear on any one of the sheets, but nothing else. Since some natural numbers are rather big, so must be the sheets. By a second step of idealization we assume that each sheet has unlimited extension, which makes it possible to write down any finite natural number whatsoever.

What do we allow our human computer, called "operator" in the sequel, to do with these sheets? Only some very simple operations. In order to describe these we shall make use of a symbolism that will

[2] See Markov [5].
[3] See Kleene [4].

reoccur throughout the remainder of this book. Let us discuss the operations one by one.

(a) →[$x_i := 0$]→. The operator writes the number zero on the sheet x_i; any prior markings on that sheet are obliterated in this process.

(b) →[$x_i := x_j$]→. The operator looks up what number is written on sheet x_j and copies it down onto sheet x_i, erasing the former content of x_i, but not changing that of x_j.

(c) →[$x_i := S(x_i)$]→.   The operator increases the number written on sheet x_i by one.

(d) →[$x_i := P(x_i)$]→.   The operator decreases the number written on sheet x_i by one; in case that number was zero, the result is zero.

Our goal is to give the operator a set of instructions, each instruction calling on the operator to perform one of the operations (a) through (d) for some indices i and j. Typically, such a set of instructions might be in the form of a diagram such as the following:

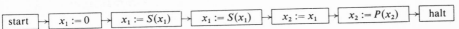

(at the end of this computation, sheet x_1 would contain the number 2 and x_2 would contain the number 1.) However, the functions computable by sequences of instructions such as the ones above form a rather meager

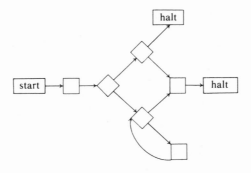

Fig. 2.1

lot [see (a) of Problem 1.8 below]. This is not overly astonishing, considering the paucity of operations and the fact that all computations must progress through the same steps, irrespective of original values and intermediate results. What is needed is the capability of *branching* in the progress of the computation. That is, our sets of instructions, or programs, should look more like the diagram shown in Fig. 2.1 than like the linear pattern employed before. This is accomplished by allowing still another capability:

(e)  · Confronted with this instruction, the operator does not change the content of x_i or any other sheet. He compares the values at x_i with zero. According to the outcome of this comparison, he progresses along one or the other of the two paths.

It remains to give precise definitions for the intuitive notion of programs above. Instead of talking about programs in terms of diagrams such as the one shown in Fig. 2.1, we take a more formal approach (but we return to diagrams presently).

1.1 Definition. Let there be given an infinite supply of variables $x_1, x_2, \ldots$, and of labels $k_1, k_2, \ldots$.

(i) The following are *operational instructions*:

$$k_p: \underline{\text{do}}\ x_i := 0\ \underline{\text{then go to}}\ k_q$$
$$k_p: \underline{\text{do}}\ x_i := x_j\ \underline{\text{then go to}}\ k_q$$
$$k_p: \underline{\text{do}}\ x_i := S(x_i)\ \underline{\text{then go to}}\ k_q$$
$$k_p: \underline{\text{do}}\ x_i := P(x_i)\ \underline{\text{then go to}}\ k_q$$
$$k_p: \underline{\text{go to}}\ k_q.$$

(ii) The following are *conditional instructions*:

$$k_p: \underline{\text{if}}\ x_i = 0\ \underline{\text{then go to}}\ k_r\ \underline{\text{else go to}}\ k_q.$$

1. THE UNIVERSAL CALCULATOR

(iii) The following are *start instructions* and *halt instructions*, respectively:

$$\text{start: } \underline{\text{go to}} \ k_q$$
$$k_p : \underline{\text{halt}}.$$

(iv) A program for the universal calculator is a finite set of instructions containing exactly one start instruction and having the property that each label k_p that occurs in the context "... $\underline{\text{go to}}\ k_p$ ..." in some instruction also occurs in the context "$k_p : ...$" in exactly one instruction.

It is easy to pass from programs to diagrams, and vice versa. An example should suffice to illustrate the procedure. We shall from now on use diagrammatic and list presentations of programs interchangeably.

1.2 Example

Program

$\quad\quad\quad\quad$ start: $\underline{\text{go to}}$ 1;
$\quad\quad\quad\quad$ 1: $\underline{\text{do}}\ x_3 := x_1\ \underline{\text{then}}\ \underline{\text{go to}}$ 2;
$\quad\quad\quad\quad$ 2: $\underline{\text{if}}\ x_2 = 0\ \underline{\text{then}}\ \underline{\text{go to}}$ 3 $\underline{\text{else}}\ \underline{\text{go to}}$ 4;
$\quad\quad\quad\quad$ 3: $\underline{\text{halt}}$;
$\quad\quad\quad\quad$ 4: $\underline{\text{do}}\ x_2 := P(x_2)\ \underline{\text{then}}\ \underline{\text{go to}}$ 5;
$\quad\quad\quad\quad$ 5: $\underline{\text{do}}\ x_3 := S(x_3)\ \underline{\text{then}}\ \underline{\text{go to}}$ 2.

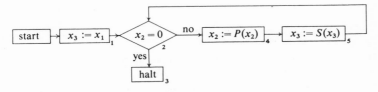

Fig. 2.2

Diagram. See Fig. 2.2. (The reader may wish to verify that the function actually computed by this program is $x_3 := x_1 + x_2$.)

While it may be intuitively clear to the reader what the computation is that is prescribed by a particular program, we still owe him a formal definition. The easiest way to understand it is to consider the actions of the operator as being taken at equal discrete time intervals, $t_1, t_2, t_3, \ldots$. At each instant t_i, we describe the present content of the sheets $x_1, \ldots, x_m$ by an m-tuple of values $a_i = \langle a_{i,1}, a_{i,2}, \ldots, a_{i,m} \rangle$. If the instruction to be performed at this instant is, say,

$$x_s := S(x_s),$$

then these values are changed accordingly, in the present case to

$$\langle a_{i,1}, \ldots, a_{i,s-1}, S(a_{i,s}), a_{i,s+1}, \ldots, a_{i,m} \rangle.$$

The pair $\langle l_i, a_i \rangle$, consisting of the label l_i of the instruction to be executed and the m-tuple a_i of values before execution, thus determines the next step. The entire history of the computation is therefore made explicit if we give the sequence of these pairs for $t_1, t_2, \ldots$. This idea is incorporated into the following definition.

1.3 Definition. Let π be a program in which only the variables $x_1, \ldots, x_m$ occur. A *computation according to* π is a finite or infinite sequence of pairs

$$\langle \underline{\text{start}}, a_0 \rangle, \langle l_1, a_1 \rangle, \ldots, \langle l_i, a_i \rangle, \langle l_{i+1}, a_{i+1} \rangle, \ldots$$

such that the following conditions are satisfied:

(i) $\underline{\text{start}}: \underline{\text{go to}}\ l_1$ is the start instruction of π;

(ii) l_{i+1} is the label of the next instruction to be executed, and a_{i+1} is the m-tuple of values obtained after the instruction with label l_i is performed on the m-tuple of values a_i;

(iii) a computation terminates if it contains a pair $\langle l_s, a_s \rangle$ where l_s: halt is an instruction in π. In this case, we call a_s the *output* of the computation for the *input* a_0.

1. THE UNIVERSAL CALCULATOR

Clearly, not all computations terminate for all inputs. For example,

start: go to 1:

1: go to 1

terminates for no input whatever, while the program shown in Fig. 2.3 terminates only for an input of even numbers. (Verify this!)

1.4 Example. See the program depicted in Fig. 2.3.

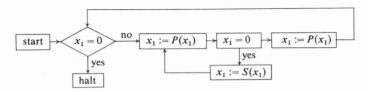

Fig. 2.3

Let us now consider a program in more detail. Suppose it is given to us in diagrammatic form.

1.5 Example. Figure 2.4 contains the diagram of a program that operates on three variables and consists of four instructions plus the start instruction. We have labeled the latter 1, 2, 3, 4, and 5.

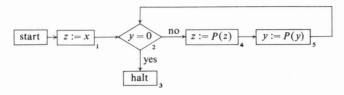

Fig. 2.4

A computation according to this program is, for example,

$$\langle \text{start}, \langle 5,3,7\rangle\rangle \quad \text{input } x = 5, y = 3, z = 7$$
$$\langle 1, \langle 5,3,7\rangle\rangle$$
$$\langle 2, \langle 5,3,5\rangle\rangle \quad \text{instruction 1 has been executed}$$
$$\langle 4, \langle 5,3,5\rangle\rangle \quad y \text{ has been found not equal to 0}$$
$$\langle 5, \langle 5,3,4\rangle\rangle \quad \text{instruction 4 has been executed}$$
$$\langle 2, \langle 5,2,4\rangle\rangle$$
$$\langle 4, \langle 5,2,4\rangle\rangle \quad y \text{ is still not equal to 0}$$
$$\langle 5, \langle 5,2,3\rangle\rangle$$
$$\langle 2, \langle 5,1,3\rangle\rangle$$
$$\langle 4, \langle 5,1,3\rangle\rangle \quad y \text{ is still not equal to 0}$$
$$\langle 5, \langle 5,1,2\rangle\rangle$$
$$\langle 2, \langle 5,0,2\rangle\rangle$$
$$\langle 3, \langle 5,0,2\rangle\rangle \quad y \text{ has been found equal to 0.}$$

Consider now any input to the program above, say $x = a_1$, $y = a_2$, $z = a_3$, abbreviated

$$\text{input} = \langle a_1, a_2, a_3 \rangle.$$

After termination of the computation, the variables x, y, z have some final values, say $x = b_1$, $y = b_2$, $z = b_3$, abbreviated

$$\text{output} = \langle b_1, b_2, b_3 \rangle.$$

The relation between input and output is given by three functions, one function for each variable, such that

$$b_1 = f_1(a_1, a_2, a_3), \quad b_2 = f_2(a_1, a_2, a_3), \quad b_3 = f_3(a_1, a_2, a_3),$$

for short

$$\langle b_1, b_2, b_3 \rangle := \langle f_1(a_1, a_2, a_3), f_2(a_1, a_2, a_3), f_3(a_1, a_2, a_3) \rangle.$$

In the present example, $f_1, f_2,$ and f_3 are *total* functions, that is, they are defined for all arguments. Indeed,

$$f_1(a_1, a_2, a_3) = a_1$$
$$f_2(a_1, a_2, a_3) = 0$$
$$f_3(a_1, a_2, a_3) = \begin{cases} a_1 - a_2 & \text{if } a_1 \geq a_2, \\ 0 & \text{otherwise.} \end{cases} = a_1 \dotminus a_2.$$

These are the functions *computed* by the program. In other examples, the program will not terminate for all inputs; the functions computed by such programs will then be *partial functions* (see Example 1.4 above).

The variables that occur in a program play quite different roles. Most often we are interested in knowing only the final value of one of the variables that is present. This would best be called the *output variable*. In Example 1.5, our choice for an output variable would be the variable z. Dually, there are those variables whose initial value really matters for our output, which in Example 1.5 would be x and y. These we call the *input variables*. A variable may at the same time be an input variable and an output variable (e.g., Example 1.4). More complicated programs have a third type of variable, auxiliary variables, whose initial values do not influence the outcome of the computation (at the output variables, at least) but are used for intermediate storage and calculations.

1.6 Example. See the program shown in Fig. 2.5. Here we have

$$\langle b_1, b_2, b_3 \rangle := \langle f_1(a_1, a_2, a_3), f_2(a_1, a_2, a_3), f_3(a_1, a_2, a_3) \rangle,$$

with

$$f_1(a_1, a_2, a_3) = a_1,$$
$$f_2(a_1, a_2, a_3) = a_1 + a_1,$$
$$f_3(a_1, a_2, a_3) = 0.$$

The only interesting outcome of the computation is that of f_2. We would call x_1 the input variable and x_2 the output variable; x_3 serves as an auxiliary variable.

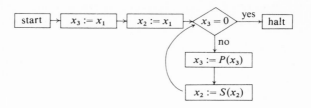

Fig. 2.5

To state the concepts more formally, let us adopt the following general definition. We have to take care to include partial functions into this definition, that is, functions which are not defined for all arguments. Such functions arise quite naturally in arithmetic, for example, the square root function (which is defined only on perfect squares) and division by 5 (defined only on multiples of 5). The counterparts of such functions, namely programs that do not terminate for all inputs, have already been encountered.

1.7 Definition. A function f on n variables is said to be *computable* by a program π with input variables $x_1, \ldots, x_n$, output variable y, and auxiliary variables $z_1, \ldots, z_m$ if the following conditions are met. If the output variable is not also an input variable, then π has exactly the variables $x_1, \ldots, x_n, y, z_1, \ldots, z_m$ and

(i) if $f(a_1,\ldots,a_n)$ is defined, then the computation according to π terminates on input $\langle a_1,\ldots,a_n,0,0,\ldots,0\rangle$ with some output $\langle b_1,\ldots,b_n,c,d_1,\ldots,d_m\rangle$ where $c = f(a_1,\ldots,a_n)$;

(ii) if π terminates on input $\langle a_1,\ldots,a_n,0,0,\ldots,0\rangle$ with output $\langle b_1,\ldots,b_n,c,d_1,\ldots,d_m\rangle$, then $f(a_1,\ldots,a_n)$ is defined and equal to c.

On the other hand, if the output variable is also one of the input variables, say x_i, then π has exactly the variables $x_1,\ldots,x_n, z_1, \ldots, z_m$ and

(i*) if $f(a_1,\ldots,a_n)$ is defined, then the computation according to π terminates on input $\langle a_1,\ldots,a_n,0,\ldots,0\rangle$ with some output $\langle b_1,\ldots,b_n,d_1,\ldots,d_m\rangle$ where $b_i = f(a_1,\ldots,a_n)$;

(ii*) if π terminates on input $\langle a_1,\ldots,a_n,0,\ldots,0\rangle$ with output $\langle b_1,\ldots,b_n,d_1,\ldots,d_m\rangle$, then $f(a_1,\ldots,a_n)$ is defined and equal to b_i.

The first goal of the present chapter is to characterize the class of functions that can be computed by programs on the universal calculator. For this purpose, we need a few tools that are to be developed in the next two sections.

1.8 Problems

(a) Consider programs in which there are no conditional instructions. Show that the functions computable by such programs are of the following form: If $a_1,\ldots,a_n$ are large enough, then

$$f(a_1,\ldots,a_n) = k \qquad \text{for some} \quad k;$$
$$f(a_1,\ldots,a_n) = a_i + k \qquad \text{for some} \quad i \quad \text{and} \quad k;$$
$$f(a_1,\ldots,a_n) = a_i \dotdiv k \qquad \text{for some} \quad i \quad \text{and} \quad k.$$

($x \dotdiv y$ is defined as $x - y$ if $x \geqslant y$ and as 0 if $x < y$.) Do we need the proviso of $a_1,\ldots,a_n$ being large enough?

(b) Consider programs that may employ all types of instructions, but which are composed in such a manner that their diagram has the form of a *tree*, as shown, for example, in Fig. 2.6. Discuss the class of functions computable by such programs. Do the problem for programs on one variable only.

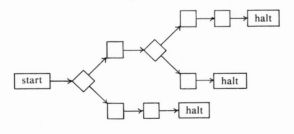

Fig. 2.6

2. EXAMPLES OF COMPUTABLE FUNCTIONS

To recognize a given (numerical) function as computable, we need to obtain a program that computes it. Now, it is not altogether necessary that the program be quite explicitly given. Indeed, very often it will suffice to observe that the function in question is defined in a certain manner from functions that we have already recognized as computable functions. In this case all we need is to parallel the combination of functions by a combination of programs.

Let us consider, for example, the function

$$f(x, y, z) = (x + y) \cdot z$$

under the assumption that we already have programs for the computation of the sum and product of two numbers. Thus we have programs that we may abbreviate to

$$\boxed{\text{start}} \to \boxed{u := x + y} \to \boxed{\text{halt}} ,$$

$$\boxed{\text{start}} \to \boxed{u := x \cdot y} \to \boxed{\text{halt}} .$$

By composing these (after conveniently renaming some variables) we obtain

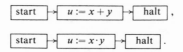

This program obviously computes the function in question (with the computed value stored at location v); u is an auxiliary variable.

More generally, consider the function f defined by composition in terms of the functions $h, g_1, \ldots, g_m$ as follows:

$$f(x_1, \ldots, x_n) = h(g_1(x_1, \ldots, x_n), \ldots, g_m(x_1, \ldots, x_n)).$$

Let us assume that we have programs

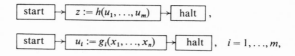

for the given functions. Then, obviously, the following program computes the function f:

start → $u_1 := g_1(x_1,\ldots,x_n)$ → ⋯ → $u_m := g_m(x_1,\ldots,x_n)$ → $z := h(u_1,\ldots,u_m)$ → halt .

Thus we have proved

2.1 Lemma. The set of computable functions is closed under composition.

Remark. Actually we have been somewhat careless in our proof. It is conceivable that during the execution of $u_1 := g_1(x_1,\ldots,x_n)$ the values of $x_1, \ldots, x_n$ are changed and that therefore u_2 already has the wrong value. We leave it to the reader to fix this up by providing some necessary memory spaces that are not unduly changed. For the future we adopt the *convention* that a program $z := f(x_1,\ldots,x_n)$ terminates always with the original values restored to the variables $x_1, \ldots, x_n$.

For another example, consider the operation of multiplication. Let us assume that we have a program which performs the operation of addition:

start → $z := x + y$ → halt .

Now, multiplication can be defined in terms of addition as we well know. Indeed it is a prime example of a definition of a function by recursion (see Chapter 1).

$$x \cdot 0 = 0,$$
$$x \cdot S(y) = x \cdot y + x.$$

It remains to convert this connection between the two operations to one between programs. One solution is depicted in Fig. 2.7. The reader can easily verify that the proposed program does compute $z := x \cdot y$, but he may be somewhat mystified as to how we went about inventing this program. This becomes clearer if we take a more general case.

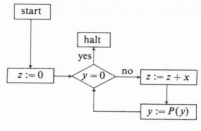

Fig. 2.7

Suppose that we are given the following programs for the functions g and h:

start → $z := g(x_1, \ldots, x_n)$ → halt ,

start → $z := h(x_1, \ldots, x_{n+2})$ → halt .

Let us consider the function f defined by the following recursion equations in terms of the given functions g and h:

$$f(x_1, \ldots, x_n, 0) = g(x_1, \ldots, x_n),$$
$$f(x_1, \ldots, x_n, S(x_{n+1})) = h(x_1, \ldots, x_n, x_{n+1}, f(x_1, \ldots, x_{n+1})).$$

Consider now the composite program shown in Fig. 2.8. This program does, indeed, calculate $z := f(x_1, \ldots, x_{n+1})$. In case $x_{n+1} = 0$, this is obvious, since the first line of the definition alone comes into play.

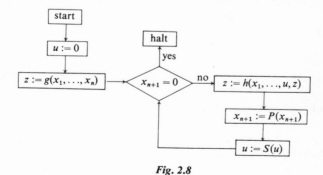

Fig. 2.8

2. EXAMPLES OF COMPUTABLE FUNCTIONS

If $x_{n+1} \neq 0$, we start going around the loop on the right-hand side, computing in succession

$$z := h(x_1, \ldots, x_n, 0, g(x_1, \ldots, x_n)) = f(x_1, \ldots, x_n, 1)$$
$$z := h(x_1, \ldots, x_n, 1, f(x_1, \ldots, x_n, 1)) = f(x_1, \ldots, x_n, 2)$$
$$\vdots$$

until we have reduced x_{n+1} to zero, which is done in x_{n+1} steps. Thus

$$\vdots$$
$$z := h(x_1, \ldots, x_n, x_{n+1} - 1, f(x_1, \ldots, x_n, x_{n+1} - 1))$$
$$:= f(x_1, \ldots, x_n, x_{n+1})$$

is computed during our last passage through the loop, and this is the output produced by the program. What we have proved, therefore, is

2.2 Lemma. The set of computable functions is closed under recursive definitions.

Or have we? What seems to have gotten lost here is the idea that g and h may be *partial* functions. But this is easily fixed up: f will still exist—as a partial function—and our proposed program will compute it. Here is the definition we need:

2.3 Definition. Let n be a nonnegative integer, let g be a partial function of n variables (a function of zero variables is simply a constant), and let h be a partial function of $n + 2$ variables. A partial function f of $n + 1$ variables is said to be determined by the recursion equations

$$f(x_1, \ldots, x_n, 0) = g(x_1, \ldots, x_n)$$
$$f(x_1, \ldots, x_n, S(x_{n+1})) = h(x_1, \ldots, x_n, x_{n+1}, f(x_1, \ldots, x_{n+1}))$$

if the following conditions are satisfied:

(i) If $f(a_1,\ldots,a_n,a_{n+1})$ is defined, then so is $f(a_1,\ldots,a_n,b)$ for all $b \leqslant a_{n+1}$, and all instances of the recursion equation corresponding to values $b \leqslant a_{n+1}$ are satisfied by f.
(ii) f is undefined in all other cases.

We say that f is "defined by primitive recursion from g and h" or that f is "determined by an application of the scheme of primitive recursion."

Let us now consider Fig. 2.9, which contains another example of a program. If we consider x as an input variable, u as output variable, and v as auxiliary variable, then the program shown in Fig. 2.9 computes the function $u := f_1(x)$. If, on the other hand, we consider u an auxiliary variable and v the output variable, then the program computes $v := f_2(x)$.

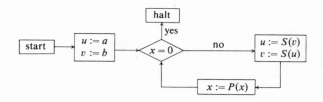

Fig. 2.9

The functions f_1 and f_2 have the following table:

x	0	1	2	3	4	...
$f_1(x)$	a	$b+1$	$a+2$	$b+3$	$a+4$	...
$f_2(x)$	b	$a+1$	$b+2$	$a+3$	$b+4$	...

These functions obviously satisfy the following equations:

$$f_1(0) = a,$$
$$f_2(0) = b,$$
$$f_1(S(x)) = S(f_2(x)),$$
$$f_2(S(x)) = S(f_1(x)).$$

2. EXAMPLES OF COMPUTABLE FUNCTIONS

Let us call these a set of *simultaneous recursion equations*. It is clear that these equations determine both f_1 and f_2. Since there is no obvious way in which we can subsume simultaneous recursion under the notion of recursion according to Definition 2.3 (see, however, Theorem 2.10*), we extend the notion of recursion equations to cover this case also.

2.4 Definition. Let n and m be nonnegative integers; let $g_1, \ldots, g_m$ be partial functions of n variables; and let $h_1, \ldots, h_m$ be partial functions of $n + m + 1$ variables. The functions $f_1, \ldots, f_m$ of $n + 1$ variables are said to be determined by the simultaneous recursion equations

$$f_j(x_1, \ldots, x_n, 0) = g_j(x_1, \ldots, x_n),$$
$$f_j(x_1, \ldots, x_n, S(x_{n+1})) = h_j(x_1, \ldots, x_n, x_{n+1}, f_1(x_1, \ldots, x_{n+1}), \ldots, f_m(x_1, \ldots, x_{n+1})),$$

$j = 1, 2, \ldots, m$, if the following conditions are satisfied:

(i) If $f_j(a_1, \ldots, a_n, a_{n+1})$ is defined, then so are all $f_i(a_1, \ldots, a_n, b)$ for $i = 1, \ldots, m$, and $b \leqslant a_{n+1}$ and all instances of the recursion equations corresponding to values $b \leqslant a_{n+1}$ are satisfied by f_i, $i = 1, \ldots, m$.
(ii) f_j is undefined in all other cases.

2.5 Lemma. The class of computable functions is closed under definitions by simultaneous recursion equations.

Proof. Suppose we are given the programs

| start | → | $y_j := g_j(x_1, \ldots, x_n)$ | → | halt |, $j = 1, \ldots, m,$

| start | → | $y_j := h_j(x_1, \ldots, x_{n+1}, y_1, \ldots, y_m)$ | → | halt |, $j = 1, \ldots, m.$

Let us consider the composite program shown in Fig. 2.10. It is easy to verify that this program computes the function f_1 with input variables $x_1, \ldots, x_n, x_{n+1}$, auxiliary variables $u, z_1, \ldots, z_m$ and $y_2, \ldots, y_m$; the output variable for f_1 is y_1. The other functions $f_2, \ldots, f_m$ are computed similarly.

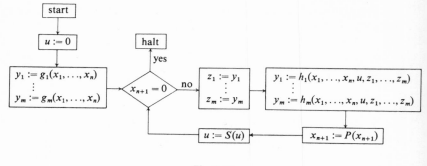

Fig. 2.10

Let us not forget that we also have some functions whose computability is established directly by programs consisting of one instruction only. To wit

2.6 Lemma. The following functions are computable:

S: the successor function, $S(x_i) = x_i + 1$;

P: the predecessor function, $P(x_i) = x_i \dotminus 1$;

Z: the function constantly equal to zero, $Z(x_i) = 0$;

U_i^n: the projection function, $U_i^n(x_1, \ldots, x_n) = x_i$;

these functions are called the *initial functions*.

Lemmas 2.1 through 2.6 already delineate an interesting class of functions, which we study in some detail later.

2.7 Definition. The class of *primitive recursive functions* is the smallest class that contains the initial functions and is closed under the schemes of composition, recursion, and simultaneous recursion.

Let us now compile a list of some well-known and useful functions that are primitive recursive (and therefore computable). In each case we indicate the type of definition employed to obtain the function in terms of functions recognized as primitive recursive earlier.

2. EXAMPLES OF COMPUTABLE FUNCTIONS

(1) *Sum*, by recursion in terms of the successor function:

$$x + 0 = x$$
$$x + S(y) = S(x + y).$$

Remark. To be quite precise, the definition above formally does not quite conform with the general scheme of recursive definition given earlier. We have tacitly used some primitive recursive functions to abbreviate the presentation, namely U_1^1 and U_3^3. In full detail, the recursive definition of addition should read

$$x + 0 = U_1^1(x)$$
$$x + S(y) = h(x, y, x + y),$$

where $h(x, y, z)$ is defined by composition as

$$h(x, y, z) = S(U_3^3(x, y, z)).$$

In the sequel, we shall, of course, suppress mention of the projection functions, thus taking advantage of the experience of the reader in reading algebraic expressions.

(2) *Product*, by recursion in terms of addition:

$$x \cdot 0 = 0$$
$$x \cdot S(y) = x \cdot y + x.$$

(3) *Power*, by recursion in terms of multiplication:

$$x^0 = 1$$
$$x^{S(y)} = x^y \cdot x.$$

(4) *Modified difference*:

$$x \dotminus 0 = x$$
$$x \dotminus S(y) = P(x \dotminus y).$$

(5) *Distance*, by composition in terms of sum and modified difference:

$$|x - y| = (x \dotminus y) + (y \dotminus x).$$

The next few functions are of technical importance in Section 4; their mathematical interest is marginal.

(6) *Signum functions*:

$$sg(0) = 0 \qquad \overline{sg}(0) = 1$$
$$sg(S(x)) = 1, \qquad \overline{sg}(S(x)) = 0.$$

(7) *Equality functions*:

$$\varepsilon(x, y) = sg(|x - y|), \qquad \bar{\varepsilon}(x, y) = \overline{sg}(|x - y|).$$

Observe that all primitive recursive functions are total functions. On the other hand, there are quite simple functions that are computable but not total. Hence the supply of computable functions is not exhausted by the primitive recursive functions. On the other hand, simple partial functions such as the one in Example 1.4 can be made primitive recursive by fixing the value of the function at some arbitrary value for arguments in which computation does not terminate. However, as we see later in the chapter, there do exist computable total functions which are not primitive recursive.

In any case, we still have to find a terminology by which we can define partial functions. For this purpose, let us consider first the example shown in Fig. 2.11. Given any value for x, this program increases the value of z starting at zero until it obtains the first z for which $z^2 = x$. In case x is not a square, the program continues increasing z indefinitely.

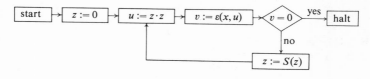

Fig. 2.11

2. EXAMPLES OF COMPUTABLE FUNCTIONS

Thus the function computed is

$$f(x) = \text{"the smallest } z, \text{ if there is one, for which } \varepsilon(x, z \cdot z) = 0,$$
$$\text{undefined otherwise."}$$

We introduce a notation for this type of definition for a function:

$$(x) = (\mu z)[\varepsilon(x, z \cdot z) = 0],$$

or, more generally, for an arbitrary function h,

$$f(x_1, \ldots, x_n) = (\mu z)[h(x_1, \ldots, x_n, z) = 0].$$

Given an arbitrary, possibly partial, function $h(x_1, \ldots, x_n, z)$ the equation above is said to define the partial function $f(x_1, \ldots, x_n)$:

$f(x_1, \ldots, x_n)$ is defined at $x_1, \ldots, x_n$ if and only if there is a number z such that $h(x_1, \ldots, x_n, y)$ is defined for all $y \leq z$ and such that z is the first number y for which the equation $h(x_1, \ldots, x_n, y) = 0$ holds.

Under these circumstances we say that f can be defined from h by an application of the μ-scheme or the scheme of minimalization.

2.8 Lemma. The class of computable functions is closed under definitions by μ-schemes.

The proof is obvious from the example above. Suppose, namely, that h is computable by the program

$$\boxed{\text{start}} \to \langle y := h(x_1, \ldots, x_n, x_{n+1}) \rangle \to \boxed{\text{halt}} \;.$$

Consider now the program shown in Fig. 2.12. This program computes the smallest z for which $h(x_1, \ldots, x_n, z) = 0$ and such that $h(x_1, \ldots, x_n, y)$ is defined for all $y \leq z$. In the case that h is not defined for some such y, or if h is never zero, the computation does not terminate.

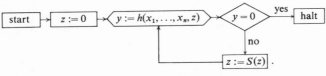

Fig. 2.12

Lemmas 2.1 through 2.8 allow us to obtain a rather large class of computable functions. We treat this class also in some detail and, therefore, give it a name.

2.9 Definition.[4] The class of *partial recursive functions* is the smallest class that contains the initial functions and is closed under the schemes of composition, recursion, simultaneous recursion, and the μ-scheme.

As we have seen, all partial recursive functions are computable (by the universal calculator); it is shown in Section 4 that the converse holds also.

A very natural question arises: Do we need all the initial functions, do we need all the schemes, for obtaining all partial recursive functions, or are some of them superfluous? The answer is that some are indeed superfluous—for example, simultaneous recursion.

2.10* Theorem. The class of primitive recursive functions is the smallest class that contains the initial functions and is closed under the schemes of composition and recursion; the class of partial recursive functions is the smallest class that is, moreover, closed under the μ-scheme.

Proof of Theorem 2.10*. Consider functions $f_1, \ldots, f_m$ defined by simultaneous recursion as follows.

$$f_i(x_1, \ldots, x_n, 0) = g_i(x_1, \ldots, x_n)$$
$$f_i(x_1, \ldots, x_n, S(x_{n+1})) = h_i(x_1, \ldots, x_{n+1}, f_1(x_1, \ldots, x_{n+1}), \ldots, f_m(x_1, \ldots, x_{n+1}))$$

where $i = 1, 2, \ldots, m$. Assume that we have defined primitive recursive functions

$$C^m, D_1^m, \ldots, D_m^m$$

without using simultaneous recursion and such that the following properties are satisfied:

---

[4] This concept of partial recursive functions is due to Kleene, who, in his pioneering paper [12] first studied the concept of total (or general) recursive function.

2. EXAMPLES OF COMPUTABLE FUNCTIONS

(i) C^m is a one-to-one onto map between the set of all m-tuples of natural numbers and the natural numbers.

(ii) $D_1^m, \ldots, D_m^m$ are the m inverse functions associated to C^m; in other words, $\ ^m(D_1^m(k), \ldots, D_m^m(k)) = k$ for all $k \geq 0$.

Let us now use these functions to define the function f by recursion as follows:

$$f(x_1, \ldots, x_n, 0) = C^m(g_1(x_1, \ldots, x_n), \ldots, g_m(x_1, \ldots, x_n))$$
$$f(x_1, \ldots, x_n, S(x_{n+1})) = C^m(h_1(x_1, \ldots, x_{n+1}, D_1^m(f(x_1, \ldots, x_{n+1})),$$
$$\vdots \qquad \ldots, D_m^m(f(x_1, \ldots, x_{n+1})))),$$
$$h_m(x_1, \ldots, x_{n+1}, D_1^m(f(x_1, \ldots, x_{n+1})),$$
$$\ldots, D_m^m(f(x_1, \ldots, x_{n+1}))))).$$

If the functions g_i and h_i are defined without use of simultaneous recursions, then our new function f is now defined without this scheme also. But observe that each function f_i can be obtained from f by

$$f_i(x_1, \ldots, x_{n+1}) = D_i^m(f(x_1, \ldots, x_{n+1}));$$

in other words, our use of simultaneous recursion for the definition of the f_i has been circumvented.

Of course, we still have to prove our assumption about the functions C^m and $D_1^m, \ldots, D_m^m$. If $m = 1$, there is obviously nothing to prove. Let us assume for the moment that we have obtained these functions for $m = 2$, namely, C^2, D_1^2, and D_2^2. The step from here to general m is simple. We let

$$C^3(x_1, x_2, x_3) = C^2(x_1, C^2(x_2, x_3)),$$
$$D_1^3(x) = D_1^2(x), \qquad D_2^3(x) = D_1^2(D_2^2(x)), \qquad D_3^3(x) = D_2^2(D_2^2(x)).$$

These functions, being defined by composition from C^2, D_1^2, and D_2^2, clearly were defined without using the scheme of simultaneous recursion.

By induction we obtain the same result for all m. It thus remains to find the three functions C^2, D_1^2, and D_2^2. Let us start with C^2, a function which is designed to enumerate pairs of natural numbers in a one-to-one fashion. Let us arrange this set of pairs in the following fashion:

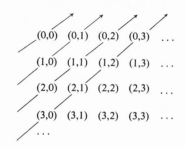

and count them according to the scheme indicated by the arrows going up the counterdiagonals. That is, we enumerate the pairs as follows:

$$(0,0), (1,0), (0,1), (2,0), (2,1), (2,2), \cdots$$

$$\underbrace{}_{\text{0th}} \underbrace{}_{\text{1st}} \underbrace{}_{\text{2nd}} \underbrace{}_{\text{3rd}}$$

$$\underbrace{}_{\text{counterdiagonals}}$$

Note that on the nth counterdiagonal we meet only pairs (i,j) for which $i + j = n$.

What is the number m associated to a pair (i,j) if we start counting at zero? Suppose (i,j) lies on the nth counterdiagonal; that is, suppose that $i + j = n$. Let us first count the number of pairs that lie on earlier counterdiagonals. This number is $1 + 2 + \cdots + n$, because on the kth counterdiagonal there are clearly just $k + 1$ pairs. On the nth counterdiagonal itself, the pair (i,j) is the $(j+1)$st element. So altogether there are $m = 1 + 2 + \cdots + n + j$ pairs which precede the pair (i,j). Recalling that $n = i + j$, we therefore see that

$$C^2(i,j) = m$$
$$= 1 + 2 + 3 + \cdots + (i+j) + j.$$

The function

$$\text{Sum}(k) = 1 + 2 + 3 + \cdots + k$$

is defined by

2. EXAMPLES OF COMPUTABLE FUNCTIONS

$$\text{Sum}(0) = 0$$
$$\text{Sum}(S(x)) = \text{Sum}(x) + S(x).$$

Thus, we obtain

$$C^2(i,j) = \text{Sum}(i+j) + j.$$

It follows by inspection that C^2 is defined without using simultaneous recursion.

We still have to show that the inverse functions D_1^2 and D_2^2 can be so defined. Consider the auxiliary function $cd(n)$ which for every n gives us the number of the counterdiagonal on which the nth pair lies. Note that n and $n+1$ lie on the same counterdiagonal iff $n+1 < C^2(cd(n)+1, 0)$. Thus

$$cd(0) = 0$$
$$cd(n+1) = cd(n) + ((n+2) \dotminus C^2(cd(n)+1, 0))$$

is a recursive definition of cd. From cd we obtain the two desired functions by composition as follows:

$$D_2^2(n) = n \dotminus C^2(cd(n), 0),$$
$$D_1^2(n) = cd(n) \dotminus D_2^2(n).$$

2.11 Problems

(a) Write a program which computes the function $\varepsilon(x,y)$.

(b) Write a program with two exits such that the first exit is taken if $x > y$, the second if $x \leqslant y$.

(c) Give a recursive definition for the factorial function and convert it into a program using only the original computational cababilities.

(d) Prove that the function $rm_p(x)$ computing the remainder of x after division by p is primitive recursive.

(e) Prove that the function

$$f(x) = \begin{cases} \sqrt{x} & \text{if } x \text{ is a square,} \\ 0 & \text{otherwise,} \end{cases}$$

is primitive recursive.

(f) Suppose that g and h are partial functions whose domains are disjoint. Then the function f defined by cases as follows

$$f(x_1,\ldots,x_n) = \begin{cases} g(x_1,\ldots,x_n) & \text{if } g(x_1,\ldots,x_n) \text{ is defined,} \\ h(x_1,\ldots,x_n) & \text{if } h(x_1,\ldots,x_n) \text{ is defined,} \\ \text{undefined otherwise} \end{cases}$$

is said to be obtained from g and h by the *disjoint union scheme*. Prove: If g and h are computable then so is f. (*Hint:* The program for f executes steps of the programs for g and h alternatively.)

2.12 Theorem. If g and h are partial recursive functions with disjoint domains then their disjoint union is a partial recursive function. (See Problem 7.3, page 160.)

3. THE STRUCTURE OF PROGRAMS

Let us consider a program π that consists of one *start instruction*, some (if any) *termination instructions*, and the remaining instructions, which form the *body of the program*. Graphically, we may represent a program as shown in Fig. 2.13, where we have replaced the three "halt" boxes by "exit 1," "exit 2," and "exit 3." This replacement is not a

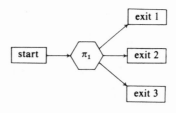

Fig. 2.13

3. THE STRUCTURE OF PROGRAMS

change in the original program but rather a notational convenience. Bodies of programs or subroutines are represented by hexagonal boxes, and individual instructions by rectangular boxes (if they are operational) and diamond-shaped boxes (if they are branching instructions).

Suppose we are given a second program, as depicted in Fig. 2.14. These two programs can be combined in many ways. It is easiest to

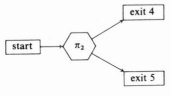

Fig. 2.14

visualize such combinations or *compositions* graphically again, by imagining that an exit of one program is identified with the start of the other. Figure 2.15 is an example of such a composition where we have given new names to the various exits of the new program. The graph suggests a composite program which still has to be formally defined, of course.

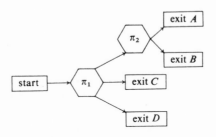

Fig. 2.15

For this purpose, let π_2' be the program obtained from π_2 by adding n to each label where n is the largest number that occurs as a label in program π_1. Suppose that p is the label of exit 1 of π_1 and that the start instruction of π_2 reads start: go to q. Then we let π_1' result from π_1 by replacing p everywhere by $q+n$. Finally, we drop the instructions $q+n$: halt from π_1' and start: go to $q+n$ from π_2'. The set $\pi_1' \cup \pi_2'$, after these deletions, is a well-formed program; it is the one represented by the diagram.

The flowchart representation makes it easy to visualize another operation on programs also, namely *looping*. Consider Fig. 2.16. The diagram in Fig. 2.17 suggests another program, which is said to have

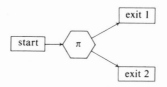

Fig. 2.16

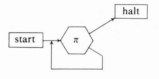

Fig. 2.17

been obtained from π by looping exit 2 back to the entrance. This new program is obtained formally from the given one by effecting the following changes: Suppose exit 2 has label p and that the start instruction reads start: go to q. All we need to do is drop the instruction p: halt, and replace p everywhere by q.

In general, a program is not just some amorphous mass of individual instructions. The most important aspect is how these instructions are combined, that is, to find its *structure*.

To discuss this structure, there are two roads open: the *synthetic* way, which consists of investigating classes of programs that are built up step by step according to some simple rules of combination; or the *analytic* way, which consists in decomposing given programs into their constituent parts.

3. THE STRUCTURE OF PROGRAMS

The goal, in any case, is to see how programs are built up, step by step, through processes such as composition and looping. In this way we can hope to discuss notions of complexity of programs and to have a systematic survey of possible programs. Such surveys are helpful in particular if we want to prove general facts about programs by some sort of mathematical induction.

Let us start by considering the most basic types of programs, namely, those whose body consists of one instruction only. We shall continue to use the symbolism introduced below to indicate the corresponding programs.

start: go to 1;
 1: do α then go to 2;
 2: halt

start: go to 1;
 1: if β then go to 2 else go to 3;
 2: halt;
 3: halt.

start: go to 1;
 1: if β then go to 2 else go to 1;
 2: halt.

start: go to 1;
 1: if β then go to 1 else go to 2;
 2: halt.

start: go to 1;
 1: if β then go to 1 else go to 1.

2. RECURSIVE FUNCTIONS AND PROGRAMMED MACHINES

For technical reasons, we also will need to consider the program whose body is empty, namely:

$$\left.\begin{array}{l}\underline{\text{start}}\colon \underline{\text{go to}}\ 1\ ;\\ 1:\underline{\text{halt}}.\end{array}\right\} \quad \boxed{\text{start}} \rightarrow \boxed{\text{halt}}$$

We are now ready to introduce one of the basic notions of this chapter, the concept of *normal form* for programs.

3.1 Definition. (i) The four programs shown in Fig. 2.18 are in normal form.

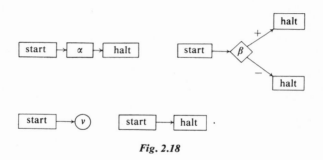

Fig. 2.18

(ii) If the program shown in Fig. 2.19 is in normal form and each program of the sort shown in Fig. 2.20 is in normal form (where we allow empty programs and programs without exits), then the composition depicted in Fig. 2.21 is in normal form.

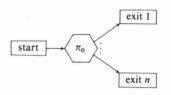

Fig. 2.19

3. THE STRUCTURE OF PROGRAMS

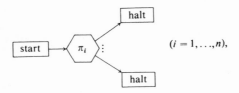

Fig. 2.20

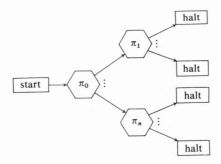

Fig. 2.21

(iii) If the program given in Fig. 2.22 is in normal form (where n may be $0, 1, \ldots$), then so is the program given in Fig. 2.23.

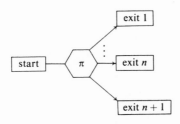

Fig. 2.22

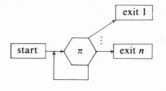

Fig. 2.23

(iv) No other programs are in normal form.

To obtain an overview of the totality of programs in normal form, let us first see what the clauses (i) and (ii) produce. Obviously, these are just the programs which are of the form of trees; see Fig. 2.24, for example. The clause (iii) adds the possibility of looping one exit back to the entrance. This looping back may be done at any time during the construction of the tree. Thus we get programs that look like the one given in Fig. 2.25. That is, all loops return to points that are located on the branch of the tree on which the loop originates. What is excluded, then, is essentially only the crossing over between branches and the recombination of branches.

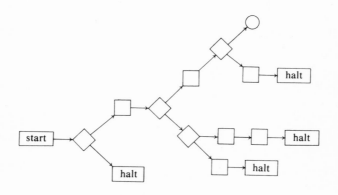

Fig. 2.24

3. THE STRUCTURE OF PROGRAMS

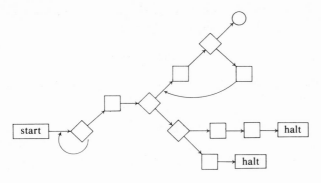

Fig. 2.25

3.2 Definition. Two *programs* are *equivalent* if they terminate on the same inputs and produce the same outputs.

3.3 Normal Form Theorem. For every program π there exists a program π' in normal form such that π' is equivalent to π.

Proof. Our proof exhibits an effective procedure which produces π', given π. We proceed by induction on the number n of instructions in the body of the program π.

The only program with $n = 0$ is

$$\boxed{\text{start}} \rightarrow \boxed{\text{halt}} \;,$$

which is normal by definition.

Induction Assumption. The theorem is true for all programs with less than n instructions in the body of the program; furthermore, to each exit of π there corresponds one or more exits of π' in such a manner that programs π and π' remain equivalent if any collection of corresponding exits are changed to nonterminating loops

$$\longrightarrow \bigcirc\!\!\!\!v$$

To continue the proof, suppose now that π has n instructions and that the nth instruction is of the form

$$k: \underline{\text{do}}\ \alpha\ \underline{\text{then}}\ \underline{\text{go}}\ \underline{\text{to}}\ p.$$

Let π_0 be the program which arises from π by replacing the instruction above by

$$k: \underline{\text{halt}}.$$

Thus π_0 has the diagram given in Fig. 2.26.

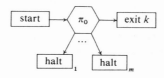

Fig. 2.26

We shall also need another program, π_p, which is obtained from π_0 by replacing the start instruction of π_0,

$$\underline{\text{start}}: \underline{\text{go}}\ \underline{\text{to}}\ \ldots,$$

by a new start instruction

$$\underline{\text{start}}: \underline{\text{go}}\ \underline{\text{to}}\ p.$$

In this way, we obtain a new program of the form shown in Fig. 2.27.

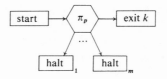

Fig. 2.27

3. THE STRUCTURE OF PROGRAMS

Now, let us compose these two programs as depicted in Fig. 2.28. By construction, π_0 and π_p have less than n instructions in their bodies, hence the induction hypothesis applies, and π_0, π_p can be replaced by equivalent programs π_0', π_p' in normal form. Then the resulting program π' (Fig. 2.29) is in normal form. It is obviously equivalent to π. The correspondence between exits of π and π' indicated by the subscripts clearly satisfies the required conditions. This finishes the case in which the added instruction is operational.

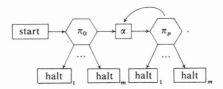

Fig. 2.28

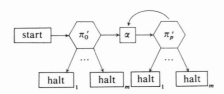

Fig. 2.29

If the added instruction is conditional, say,

$$k: \underline{\text{if }} \alpha \underline{\text{ then go to }} p \underline{\text{ else go to }} q,$$

we let π_0, π_p, and π_q be constructed as above and composed as shown in Fig. 2.30. The rest of the argument is the same, finishing the proof.

124   2. RECURSIVE FUNCTIONS AND PROGRAMMED MACHINES

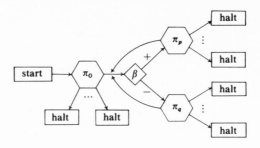

Fig. 2.30

Remarks. To be quite precise, we have to distinguish the various instances of instructions "... go to k ..." which occur in π_0 and keep them separate. Thus, π_0 is really of the form given in Fig. 2.31 and the composite program should, in the first case, look like the one shown in

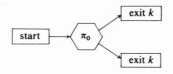

Fig. 2.31

Fig. 2.32. Otherwise, the composed program will not, in general, be in normal form. (Why?)

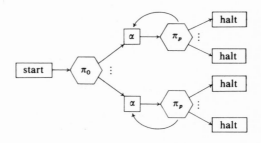

Fig. 2.32

3. THE STRUCTURE OF PROGRAMS

3.4 Problems

(a) Show that the following programs are (or are not) in normal form, illustrating thereby that you have a decision procedure for this question.

start: go to 1
1: if β_1 then go to 2 else go to 3;
2: if β_2 then go to 7 else go to 8;
3: do α_1 then go to 4;
4: if β_1 then go to 5 else go to 9;
5: if β_2 then go to 3 else go to 6;
6: do α_2 then go to 4;
7: halt;
8: do α_1 then go to 8;
9: halt.

start: go to 1
1: if β_1 then go to 2 else go to 3;
2: do α_1 then go to 5;
3: do α_2 then go to 4;
4: if β_2 then go to 3 else go to 5;
5: if β_1 then go to 6 else go to 7;
6: do α_1 then go to 3;
7: halt.

(b) Find a program in normal form which is equivalent to that program in problem (a) which is not in normal form.

Solution. The diagram of the program in question is given in Fig. 2.33. We decide to cut out instruction 2, thereby we obtain π_0 (Fig. 2.34), and π_5 (Fig. 2.35), or, equivalently, deleting the inaccessible instruction on the left, the form of π_5 shown in Fig. 2.36. Fortunately both π_0 and π_5 are already in normal form (otherwise we would have to go on, cutting out an instruction in π_0, one in π_5, and repeat the procedure until we

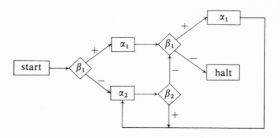

Fig. 2.33

obtain programlets in normal form). As it is we can now go to the next step in our algorithm and paste π_0 and (the modified) π_5 together to obtain the diagram shown in Fig. 2.37. The resulting program is depicted in Fig. 2.38.

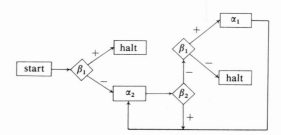

Fig. 2.34

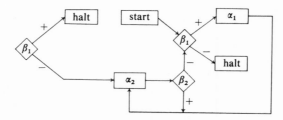

Fig. 2.35

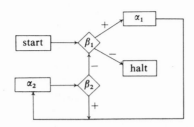

Fig. 2.36

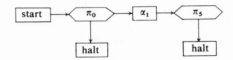

Fig. 2.37

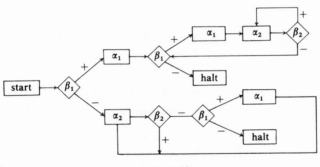

Fig. 2.38

4. COMPUTABLE FUNCTIONS

A very straightforward application of the normal form theorem gives us a proof of the fact that all functions computable on the universal calculator are partial recursive. By Definition 1.7, this amounts to proving the following.

4.1 Theorem. Let π be a program whose variables are in the set $\{x_1,\ldots,x_n\}$, and let f_j be the function computed by π with output variable x_j, $j \leq n$. Then f_j is a partial recursive function.

Proof. To simplify matters we consider all variables $x_1, \ldots, x_n$ (even those that do not occur in π) as input variables. This can be easily accomplished by providing π with some additional but inactive instructions. Hence the functions f_j all depend on n variables. If these functions are partial recursive, then clearly so are the original ones, that is, the functions computed by π in the sense of Definition 1.7. (Why?) We may assume that π is in normal form (a normal form can, in any case, be effectively obtained). We proceed by induction on the structure of π as follows.

Induction Assumption. For each $j = 1, 2, \ldots, n$, the function f_j is partial recursive; moreover, the characteristic functions g_i of the exits i, defined by

$$g_i(a_1,\ldots,a_n) = \begin{cases} 0 & \text{if } \pi \text{ takes exit } i \text{ on input } \langle a_1,\ldots,a_n \rangle, \\ 1 & \text{if } \pi \text{ takes some other exit on this input,} \\ \text{undefined otherwise,} \end{cases}$$

are each partial recursive.

The induction assumption is easily verified for the four miniature programs in clause (i) of Definition 3.1:

For

4. COMPUTABLE FUNCTIONS

we have the following cases:

α	$f_j(x_1, \ldots, x_n)$	$g(x_1, \ldots, x_n)$
$x_i := S(x_i)$	$S(U_i^n)$	$Z(U_1^n)$
$x_i := P(x_i)$	$P(U_i^n(x_1, \ldots, x_n))$	$Z(U_1^n)$
$x_i := x_p$	$U_j^n(U_1^n, \ldots, U_{i-1}^n, U_p^n, \ldots, U_n^n)$	$Z(U_1^n)$
$x_i := 0$	$U_j^n(U_1^n, \ldots, U_{i-1}^n, Z, \ldots, U_n^n)$	$Z(U_1^n)$.

For

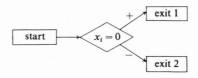

we have

$$f_j(x_1, \ldots, x_n) = U_j^n$$
$$g_1(x_1, \ldots, x_n) = sg(U_1^n)$$
$$g_2(x_1, \ldots, x_n) = \overline{sg}(U_1^n).$$

For

start → halt

we have

$$f_j(x_1, \ldots, x_n) = U_j^n$$
$$g(x_1, \ldots, x_n) = Z(U_1^n).$$

Finally, for

start → ⓥ

we have

$$f_j(x_1, \ldots, x_n) = (\mu x)[S(x) = 0].$$

Obviously, all the functions f_j and g_k introduced above are partial recursive.

For the induction step we need to consider composition and looping. It is easy to see that we may restrict attention to two simple cases of composition, namely, those given in Figs. 2.39 and 2.40. The type of

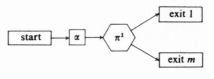

Fig. 2.39

composition shown in Fig. 2.39 is handled by cases as follows (where f_j' and g_k' are the functions associated to the program π^1 by induction assumption):

α	$f_j(x_1,\ldots,x_n)$	$g_k(x_1,\ldots,x_n)$
$x_i := S(x_i)$	$f_j'(U_1^n,\ldots,U_{i-1}^n,S(U_i^n),\ldots,U_k^n)$	$g_k'(U_1^n,\ldots,U_{i-1}^n,S(U_i^n),\ldots,U_n^n)$
$x_i := P(x_i)$	$f_j'(U_1^n,\ldots,U_{i-1}^n,P(U_i^n),\ldots,U_n^n)$	$g_k'(U_1^n,\ldots,U_{i-1}^n,P(U_i^n),\ldots,U_n^n)$
$x_i := x_p$	$f_j'(U_1^n,\ldots,U_{i-1}^n,U_p^n,\ldots,U_n^n)$	$g_k'(U_1^n,\ldots,U_{i-1}^n,U_p^n,\ldots,U_n^n)$
$x_i := 0$	$f_j'(U_1^n,\ldots,U_{i-1}^n,Z,\ldots,U_n^n)$	$g_k'(U_1^n,\ldots,U_{i-1}^n,Z,\ldots,U_n^n)$

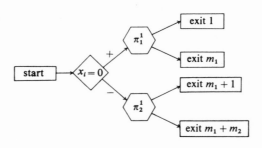

Fig. 2.40

To treat the second type of composition (Fig. 2.40), assume that f_j' and f_j'' are the partial recursive functions computed by π_1 and π_2, respectively, and let g'_k and g''_k be the characteristic functions of the exits of these programs. The function f_j computed by the composite program is given by the disjoint union scheme

$$f_j(x_1,\ldots,x_n) = \begin{cases} f'_j(x_1,\ldots,x_n) \cdot (\mu y)[x_i + P(y) = 0] \text{ if defined,} \\ f''_j(x_1,\ldots,x_n) \cdot (\mu y)[P(x_i) + P(y) + \overline{sg}(x_i) = 0] \text{ if defined,} \\ \text{undefined otherwise.} \end{cases}$$

and is, by Theorem 2.12, partial recursive. To determine the characteristic functions of the exits of the composite program consider first the case $0 < k \leq m_1$. The following disjoint union scheme defines g_k:

$$g_k(x_1,\ldots,x_n) = \begin{cases} g'_k(x_1,\ldots,x_n) \cdot (\mu y)[x_1 + P(y) = 0] \text{ if defined,} \\ P(g''_1(x_1,\ldots,x_n)) + (\mu y)[P(x_i) + P(y) + \overline{sg}(x_i) = 0] \text{ if defined,} \\ \text{undefined otherwise.} \end{cases}$$

The case $m_1 < k \leq m_1 + m_2$ is treated analogously. In either case the functions g_k are clearly partial recursive.

Next, let us consider the case of looping one exit of the program shown in Fig. 2.41, which results in the program shown in Fig. 2.42. Let π contain the variables $x_1, \ldots, x_n$ exclusively, and let p_j be the functions computed by the program depicted in Fig. 2.43 with input variables $x_1, \ldots, x_n$ and output variable x_j, $j = 1, \ldots, n$. Similarly, let q_j be the functions computed by the program shown in Fig. 2.44 with input variables $x_1, \ldots, x_n$ and output variable x_j, $j = 1, \ldots, n$. By what we just proved, the functions p_j and q_j are partial recursive functions.

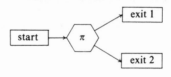

Fig. 2.41

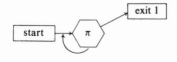

Fig. 2.42

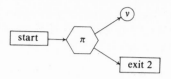

Fig. 2.43

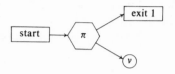

Fig. 2.44

Let $p_j^{(y)}$ be the y-fold iteration of p_j. It is defined by recursion as follows [writing $h_j(y; x_1, \ldots, x_n)$ for $p_j^{(y)}(x_1, \ldots, x_n)$]:

$$h_j(0; x_1, \ldots, x_n) = x_j$$
$$h_j(S(y); x_1, \ldots, x_n) = p_j(h_1(y; x_1, \ldots, x_n), \ldots, h_n(y; x_1, \ldots, x_n)).$$

Note that we are defining here n functions simultaneously since each h_i is involved in the two defining equations of h_j. By Definition 2.9, this procedure does not lead us outside the family of partial recursive functions.

Let us now define the function k by

$$k(x_1, \ldots, x_n) = (\mu y)[g_1(h_1(y; x_1, \ldots, x_n), \ldots, h_n(y; x_1, \ldots, x_n)) = 0],$$

where g_1 is the characteristic function of exit 1. Observe that k gives us the least number of times which the loop has to be run through before the next execution of π leads us to exit 1. The values in storage at $x_1, \ldots, x_n$ at this moment are

$$h_1(k(x_1, \ldots, x_n); x_1, \ldots, x_n), \ldots, h_n(k(x_1, \ldots, x_n); x_1, \ldots, x_n),$$

respectively. The program, therefore, computes

$$f_j(x_1, \ldots, x_n) = q_j(h_1(k(x_1, \ldots, x_n); x_1, \ldots, x_n), \ldots, h_n(k(x_1, \ldots, x_n); x_1, \ldots, x_n)),$$

obviously a partial recursive function, $j = 1, \ldots, n$.

It remains for us to observe that the characteristic function for the one exit of the looped program can be represented by

$$g(x_1, \ldots, x_n) = Z(f_1(x_1, \ldots, x_n)),$$

since the right-hand side of this equation is zero when f is defined, that is, when the exit is taken, undefined otherwise. In the more general case of item (iii) of Definition 3.1, the characteristic functions are obtained rather easily also.

The generality of the proof above may perhaps hide the very simple idea that is behind it. To bring this idea out more clearly, let us consider an example.

4.2 Example. Consider the program shown in Fig. 2.45. It is

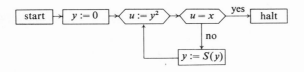

Fig. 2.45

obviously in normal form. Let us find the partial recursive function computed by this program. The method of proof for Theorem 4.1 indicates how we should go about this.

As a first step, we need to analyze the structure of the program (Fig. 2.45) by a sequence of programs in normal form of increasing complexity. The functions computed by the first few programs in this sequence and the characteristic functions are found immediately; the corresponding functions for later programs are then built up in a systematic way from these.

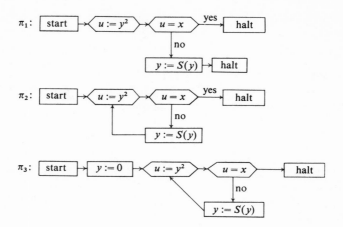

Fig. 2.46

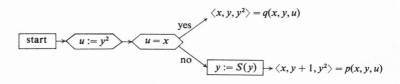

Fig. 2.47

In the case at hand, the analysis of the program gives us the sequence of programs depicted in Fig. 2.46. Let us consider π_1 first, as augmented in Fig. 2.47 where we indicate the outputs at the two exits as vectors in convenient notation. The characteristic functions of the exits are, respectively,

$$g_1(x,y,u) = \begin{cases} 0 & \text{if } x = y^2 \\ 1 & \text{otherwise,} \end{cases} \quad g_2(x,y,u) = \begin{cases} 1 & \text{if } x = y^2 \\ 0 & \text{otherwise;} \end{cases}$$

that is, $g_1(x,y,u) = \varepsilon(x, y^2)$ and $g_2(x,y,u) = \bar{\varepsilon}(x, y^2)$.

4. COMPUTABLE FUNCTIONS

Now, consider π_2. The (vector) function f computed by π_2 is found as follows. Let

$$h(z; x, y, u) = p^{(z)}(x, y, u) = \langle x, y + z, (y + z - 1)^2 \rangle$$

and

$$\begin{aligned}k(x, y, u) &= (\mu z)[g_1(h(z; x, y, u)) = 0] \\ &= (\mu z)[g_1(x, y + z, (y + z \dot{-} 1)^2) = 0] \\ &= (\mu z)[\varepsilon(x, (y + z)^2) = 0].\end{aligned}$$

Then

$$\begin{aligned}f(x, y, u) &= q(h(k(x, y, u); x, y, u)) \\ &= q(h((\mu z)[\varepsilon(x, (y + z)^2) = 0]; x, y, u)) \\ &= q(x, y + (\mu z)[\varepsilon(x, (y + z)^2) = 0], (y + (\mu z)[\cdots] \dot{-} 1)^2) \\ &= \langle x, y + (\mu z)[\varepsilon(x, (y + z)^2) = 0], (y + (\mu z)[\cdots])^2 \rangle.\end{aligned}$$

Finally, the function $F(x, y, u)$ computed by the original program, which is

$$\boxed{\text{start}} \rightarrow \boxed{y := 0} \rightarrow \langle \pi_2 \rangle \rightarrow \boxed{\text{halt}},$$

is found by compositions as follows:

$$\begin{aligned}F(x, y, u) &= f(x, 0, u) \\ &= \langle x, (\mu z)[\varepsilon(x, z^2) = 0], ((\mu z)[\varepsilon(x, z^2) = 0])^2 \rangle.\end{aligned}$$

The output value in which we are interested is that at y. According to the calculations above, this is represented by the function

$$y := (\mu z)[\varepsilon(x, z^2) = 0].$$

or, by abuse of notation,

$$y := (\mu z)[x = z^2] = \begin{cases} +\sqrt{x} & \text{if it exists,} \\ \text{undefined} & \text{otherwise.} \end{cases}$$

136   2. RECURSIVE FUNCTIONS AND PROGRAMMED MACHINES

Of course, this could have been guessed directly from looking at the program, since that is so very simple. But, especially for more involved programs, it is good to have a general procedure.

4.3 Problems
(a) Find the function computed by the program shown in Fig. 2.48.

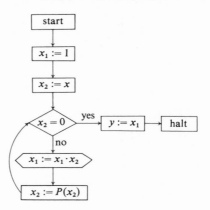

Fig. 2.48

(b) Do the same for the program shown in Fig. 2.49.

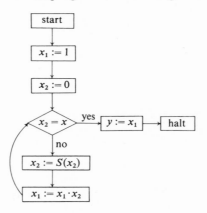

Fig. 2.49

(c) Suppose that $g_1, \ldots, g_n$ and $g'_1, \ldots, g'_m$ are the characteristic functions of the exits of π_1 and π_2, respectively. Find the characteristic functions of π_1 composed with π_2 (at any one of the exits of π_1).

(d) Suppose that $g_1, \ldots, g_n, g_n + 1$ are the characteristic functions of the exits of π, and f is the function computed by π. Find the characteristic functions $g'_1, \ldots, g'_n$ of the exits 1 to n if exit $n + 1$ is looped back to the entrance.

5. LOOP PROGRAMS AND PRIMITIVE RECURSIVE FUNCTIONS

Let us now return to the class of primitive recursive functions. A quick check on the definition of this class (as simplified by Theorem 2.10*) reveals that each function in the class is total; indeed we can very easily convince ourselves that a bound on the number of computation steps is available for each argument. Namely, let us agree that the functions P, S, Z, and U_i^n use but one step in their computation. Now, if we know how many steps it takes to compute $g_1, \ldots, g_m$ and h, then we know it for the composite $h(g_1, \ldots, g_m)$. Also, if we know the facts for g and h, then the function defined by recursion,

$$f(x_1, \ldots, x_n, 0) = g(x_1, \ldots, x_n)$$
$$f(x_1, \ldots, x_n, S(x_{n+1})) = h(x_1, \ldots, x_{n+1}, f(x_1, \ldots, x_{n+1})),$$

allows an estimation; indeed in the computation of $f(x_1, \ldots, x_{n+1})$, the second equation is used x_{n+1} times, and the first once.

On the other hand, a program for an arbitrary partial recursive function does not admit of such an easy estimate. For example, it is easy to write a program for the function f: $f(n)$ equals the smallest number k, if there is one, such that there exist x, y, and z less than k with $x^n + y^n = z^n$, undefined otherwise. The reader is invited to provide this program [(c) of Problem 5.3]. He will see that the main obstacle to an estimate is the fact that this program has a loop about which we have no idea how often it will be run through. (If we had, we would know a lot more about Fermat's last problem: Are there integer solutions of $x^n + y^n = z^n$ for all $n \geqslant 3$?)

138   2. RECURSIVE FUNCTIONS AND PROGRAMMED MACHINES

The conclusion, then, is that, for programs that compute primitive recursive function, we do have an idea how often a loop is run through. Let us therefore make this an express feature of a class of programs and hope for the best, namely, that each primitive recursive function is computable by a program in this class, and vice versa. Let us take a program

$$\boxed{\text{start}} \to \langle \pi \rangle \to \boxed{\text{halt}}$$

which does not contain the variable y and consider the program

<u>start</u>: <u>go to</u> 1;
   1: <u>if</u> $y = 0$ <u>then go to</u> 4 <u>else go to</u> 2;
   2: <u>do</u> $y := P(y)$ <u>then go to</u> 3;
   3: <u>do</u> π <u>then go to</u> 1;
   4: <u>halt</u>.

Here, we have used the abbreviation "3: <u>do</u> π <u>then go to</u> 1" to indicate the obvious composition of programs. We use the diagram given in Fig. 2.50 to present this composite program. It is important that the reader keeps in mind that if y happens to be zero, the portion $\to \langle \pi \rangle \to$ is bypassed; in case it is $k > 0$, it is run through exactly k times.

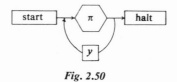

Fig. 2.50

5.1 Definition.[5] Let $L_0 \subseteq L_1 \subseteq L_2 \subseteq \cdots$ be the following sets of programs:

(i) L_0 consists of all programs which contain only instructions of the form $x_i := x_j$, $x_i := S(x_i)$, and $x_i := 0$.

---

[5] The concept and theory of loop programs are due to Meyer and Ritchie [13].

5. LOOP PROGRAMS AND PRIMITIVE RECURSIVE FUNCTIONS

(ii) If $\pi \in L_{n-1}$ and y does not occur in π, then the program shown in Fig. 2.51 is in L_n.

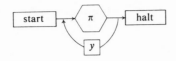

Fig. 2.51

(iii) If π is in L_{n-1}, then π is in L_n; if π_1 and π_2 are in L_n, then so is

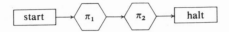

The programs in $L = \bigcup_{n=0}^{\infty} L_n$ are called *loop programs*. If $\pi \in L_n - L_{n-1}$, then n is called the *loop complexity* of π. Observe that all loop programs have exactly one exit (which is always taken).

Example. The loop program shown in Fig. 2.52 computes the value $x + y$ at x.

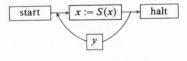

Fig. 2.52

5.2 Theorem. A function is primitive recursive if and only if it is computable by a loop program.

Proof. Let f be any primitive recursive function of n variables. We shall show that there is a loop program π_f containing variables $x_1, \ldots, x_n$ and y such that whenever π_f is started with $y = 0$, it terminates with

$y = f(x_1, \ldots, x_n)$, and the original values of the input are restored to the variables x_i. To indicate this situation, we write this program as

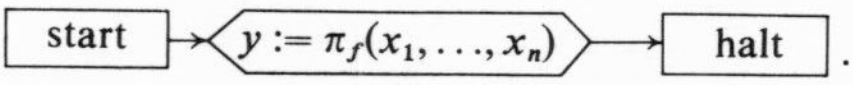

$$\boxed{\text{start}} \to \langle y := \pi_f(x_1, \ldots, x_n) \rangle \to \boxed{\text{halt}}.$$

(a) *Successor function S*:

> start: go to 1:
> 1: do $y := S(x)$ then go to 2;
> 2: halt.

(b) *Projection functions U_i^n*:

> start: go to 1;
> 1: do $y := x_i$ then go to 2;
> 2: halt.

(c) *Zero function Z*:

> start: go to 1;
> 1: do $y := 0$ then go to 2;
> 2: halt.

(d) *Composition.* Suppose that h is m-ary, and that g_i are n-ary primitive recursive functions, $i = 1, \ldots, m$. Let π_h and π_{g_i} be the corresponding loop programs. Then the program shown in Fig. 2.53 computes $y = h(g_1(x_1, \ldots, x_n), \ldots, g_m(x_1, \ldots, x_n))$.

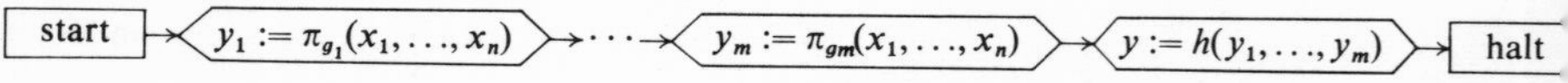

$$\boxed{\text{start}} \to \langle y_1 := \pi_{g_1}(x_1, \ldots, x_n) \rangle \to \cdots \to \langle y_m := \pi_{gm}(x_1, \ldots, x_n) \rangle \to \langle y := h(y_1, \ldots, y_m) \rangle \to \boxed{\text{halt}}$$

Fig. 2.53

(e) *Recursion.* Suppose that g and h are $(n-1)$-ary, respectively, $(n+1)$-ary primitive recursive functions computed by loop programs π_g and π_h, respectively. Let f be the n-ary function defined by

5. LOOP PROGRAMS AND PRIMITIVE RECURSIVE FUNCTIONS

$$f(x_1, \ldots, x_{n-1}, 0) = g(x_1, \ldots, x_{n-1})$$
$$f(x_1, \ldots, x_{n-1}, S(x_n)) = h(f(x_1, \ldots, x_n), x_1, \ldots, x_n).$$

Then the program shown in Fig. 2.54 obviously computes $y := f(x_1, \ldots, x_n)$. By Theorem 2.10* the cases (a)–(e) are all that need to be checked.

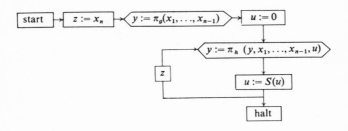

Fig. 2.54

To prove the converse is just as straightforward: Clearly, all loop programs with just one instruction compute primitive recursive functions. If π_1 computes f and π_2 computes g, then their composition

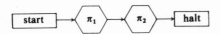

computes $g(f)$, which is (coordinatewise) a primitive recursive function. Finally, if

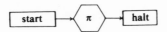

computes

$$g_1(x_1, \ldots, x_n), \ldots, g_n(x_1, \ldots, x_n)$$

for output variables $x_1, \ldots, x_n$, respectively, then the program shown in Fig. 2.55 computes the functions $f_1^{(z)}, \ldots, f_n^{(z)}$ at these locations, where

$$f_i^{(0)}(x_1,\ldots,x_n) = x_i,$$
$$f_i^{(S(z))}(x_1,\ldots,x_n) = g_i(f_1^{(z)}(x_1,\ldots,x_n),\ldots,f_n^{(z)}(x_1,\ldots,x_n)),$$

for $i = 1, \ldots, n$. These functions, being thus defined by simultaneous recursion from primitive recursive functions, are themselves primitive recursive. The function computed for output variable z is the zero function and therefore also primitive recursive.

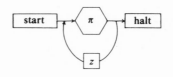

Fig. 2.55

5.3 Problems

(a) Write a loop program that computes $x \cdot y$ at z.

(b) What function $f(x, y)$ is computed by the program given in Fig. 2.56 at x?

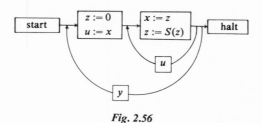

Fig. 2.56

(c) Write a program for the function $f(n)$ defined at the start of this section.

(d) Write a loop program for the predecessor function P, for the modified difference, and for the factorial.

(e) Show that $\max(x_1,\ldots,x_m)$ which computes the maximum of $x_1, \ldots, x_m$ is a primitive recursive function by obtaining a loop program for it.

(f)* Let g be given by a loop program. Consider the function $f(x,y)$ defined by

$$f(x,y) = \mu_{z=0}^{y}[g(x,z) = 0] = \begin{cases} \text{smallest} \quad z, \quad 0 \leq z \leq y, \\ \text{such that} \quad g(x,z) = 0 \quad \text{if such exist;} \\ 0 \text{ otherwise.} \end{cases}$$

Find a loop program for f.

6. *COMPLEXITY AND GROWTH OF PRIMITIVE RECURSIVE FUNCTIONS

Let g be a primitive recursive function computed by the program π. Suppose that $\pi \in L_n$ but $\pi \notin L_{n-1}$; that is, suppose that π has loop complexity n. What can we say offhand about the function g?

One kind of statement that we may wish to make is about the growth rate of g. For this we adopt the following terminology.

6.1 Definition. Let g be an m-ary function, f a unary function.

(a) We say that g is *bounded* by f if $g(x_1,\ldots,x_n) < f(\max(x_1,\ldots,x_n))$ for all $x_1, \ldots, x_n$.

(b) We say that g is *majorized* by f if there exists N such that $g(x_1,\ldots,x_n) < f(\max(x_1,\ldots,x_n))$ for all $x_1,\ldots,x_n$ such that

$$\max(x_1,\ldots,x_n) \geq N.$$

Our goal is to obtain a sequence of unary functions $f_0, f_1, f_2, \ldots$ such that

(i) $f_{k+1} \in L_{k+1} - L_k$,
(ii) if $g \in L_k$, then g is majorized by f_{k+1}.

Thus, in particular, f_{k+1} majorizes f_k, and we have a sequence of faster and faster growing functions.

In order to define the functions f_k, we make use of the process of iteration.

6.2 Definition. Suppose that g is a unary function; let h be the binary function defined by recursion as follows:

$$h(x, 0) = x$$
$$h(x, S(y)) = g(h(x, y)).$$

We use the symbolism $g^{(y)}(x)$ to indicate this function $h(x, y)$ and read it as "the y-fold iterate of g at x."

Now we can proceed to present the functions f_k.

6.3 Definition. Let $f_0, f_1, \ldots$ be the functions defined by

$$f_0(x) = \begin{cases} 1 & \text{if } x = 0, \\ 2 & \text{if } x = 1, \\ x + 2 & \text{if } x > 1; \end{cases}$$

$$f_{n+1}(x) = f_n^{(x)}(1).$$

Let us now collect some easy observations about the functions f_k which we prove by induction.

(1) $f_1(x) = 2x + (1 \dotminus x)$.

Proof

$$f_1(0) = f_0^{(0)}(1) = 1 = 2 \cdot 0 + (1 \dotminus 0) = 1.$$
$$f_1(1) = f_0^{(1)}(1) = f_0(1) = 2 = 2 \cdot 1 + (1 \dotminus 1).$$
$$f_1(x+1) = f_0^{(x+1)}(1) = f_0(f_0^{(x)}(1)) = f_0^{(x)}(1) + 2 = 2 \cdot x + (1 \dotminus x) + 2$$
$$= 2(x+1) + (1 \dotminus x) = 2(x+1) + (1 \dotminus (x+1)) \qquad \text{if } x > 0.$$

(2) $f_1^{(p+1)}(x) = 2^p \cdot f_1(x) \geq 2^{p+1} \cdot x$.

Proof:

$$f_1^{(1)}(x) = f_1(x) = 2^0 \cdot f_1(x) = 2x + (1 \dotminus x) \geq 2x = 2^1 \cdot x.$$

6. COMPLEXITY AND GROWTH OF PRIMITIVE RECURSIVE FUNCTIONS

$$\begin{aligned}
f_1^{(p+1+1)}(x) = f_1(f_1^{(p+1)}(x)) &= f_1(2^p \cdot f_1(x)) \\
&= 2 \cdot (2^p \cdot f_1(x)) + (1 \dotdiv 2^p \cdot f_1(x)) \quad \text{by induction hypothesis,} \\
&= 2^{p+1} \cdot f_1(x) \quad \text{since } f_1(x) \geq 1 \\
&\qquad \text{by (1) and hence} \quad 1 \dotdiv 2^p \cdot f(x) = 0, \\
&\geq 2^{p+1} \cdot 2 \cdot x \quad \text{since } f_1(x) \geq 2x \quad \text{by (1),} \\
&= 2^{(p+1)+1} \cdot x.
\end{aligned}$$

(3) $f_2(x) = 2^x$.

Proof

$$f_2(0) = f_1^{(0)}(1) = 1 = 2^0.$$
$$f_2(p+1) = f_1^{(p+1)}(1) = 2^p \cdot f_1(1) = 2^p \cdot f_1(1) = 2^p \cdot 2 = 2^{p+1}.$$

(4) $f_n(x) \geq x + 1$.

Proof

$$\begin{aligned}
f_0(x) &= x + 1 \quad \text{for } x = 0, 1. \\
f_0(x) &= x + 2 > x + 1 \quad \text{for } x > 1. \\
f_{n+1}(x) &= f_n^{(x)}(1) \geq f_n(1) + \underbrace{1 + 1 + \cdots + 1}_{x} \geq x + 1.
\end{aligned}$$

(5) $f_n^{(p)}(x) \leq f_n^{(p)}(x+1), f_n^{(p)}(x) < f_n^{(p+1)}(x), f_n^{(p)}(x) \leq f_{n+1}^{(p)}(x)$.

(6) If $n \geq 1$, then $f_n^{(p+1)}(x) \geq f_n^{(p)}(x) \cdot 2$.

Proof

$$f_1^{(p+1)}(x) = f_1(f_1^{(p)}(x)) = 2 \cdot f_1^{(p)}(x) + (1 \dotdiv f_1^{(p)}(x)) \geq f_1^{(p)}(x) \cdot 2.$$
$$f_{n+1}^{(p+1)}(x) = f_{n+1}(f_{n+1}^{(p)}(x)) \geq f_1(f_{n+1}^{(p)}(x)) \geq 2 \cdot f_{n+1}^{(p)}(x).$$

(7) If $n \geq 2$, then $(f_n^{(p)}(x))^2 \leq f_n^{(p+2)}(x)$.

The proofs of (5) and (7) are left to the reader [item (a) of Problem 6.8].

We are now prepared to tackle the question of growth rate. Let us first remark about a connection between the *value* $g_\pi(a_1,\ldots,a_n)$ of the function g_π computed by π for input $\langle a_1,\ldots,a_n\rangle$ and the *number of steps* $t_\pi(a_1,\ldots,a_n)$ this computation takes. Suppose that t_π is bounded by f. Then g_π is bounded by h where $h(y) = f(y) + y$. This is obvious since each computation step increases the value of at most one x_i by at most 1. Thus, it is sufficient to obtain bounds for the functions t_π.

6.4 Theorem. For every loop program of loop complexity n, we can effectively find a number p such that t_π is bounded by $f_n^{(p)}$.

Corollary. The function g_π computed by a loop program of loop complexity n is bounded by $f_n^{(p')}$ for some p'. Namely,

$$g_\pi(a_1,\ldots,a_k) \leq t_\pi(a_1,\ldots,a_k) + \max(a_1,\ldots,a_k)$$
$$< f_n^{(p)}(\max(a_1,\ldots,a_k)) + \max(a_1,\ldots,a_k)$$
$$\leq f_n^{(p+1)}(\max(a_1,\ldots,a_k)) \qquad \text{by (4) and (6).}$$

Proof of Theorem 6.4. (We give an outline only; the reader can easily fill in the necessary estimates using observations (1)–(7) above.) Let $m = \max(a_1,\ldots,a_k)$, and let q be the number of instructions in π_0. If $\pi_0 \in L_0$, then π_0 has no loop and

$$t_{\pi_0}(a_1,\ldots,a_k) \leq q < f_0^{(q)}(0) \leq f_0^{(q)}(m).$$

If $\pi \in L_1$ is obtained as

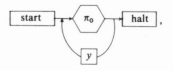

then
$$t_\pi(a_1,\ldots,a_k) \leq (q+1)\cdot m + 2 \leq 2^q \cdot m + 2 \leq f_1^{(q)}(m) + 2 \leq f_1^{(q+2)}(m).$$

Now let $\pi_0 \in L_{n-1}$ and assume that

$$t_{\pi_0}(a_1, \ldots, a_{k-1}) \leq f_{n-1}^{(q)}(m),$$

where $m = \max(a_1, \ldots, a_{k-1})$. Consider $\pi \in L_n$ obtained as

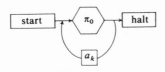

with $\pi_0 \in L_{n-1}$. Now, after the first execution of π_0, the contents of x_i, $i = 1, \ldots, k-1$, are bounded by $m + f_{n-1}^{(q)}(m) \leq f_{n-1}^{(q+1)}(m)$, since each step can increase by at most 1 the value of any x_i. Thus second executions will take at most $f_{n-1}^{(q)}(f_{n-1}^{(q+1)}(m))$ steps and will leave the x_i's with values bounded by

$$f_{n-1}^{(q)}(f_{n-1}^{(q+1)}(m)) + f_{n-1}^{(q+1)}(m) \leq f_{n-1}^{(2(q+1))}(m).$$

By induction, the ith repetition of the loop will take at most $f_{n-1}^{(i(q+1))}(m)$ steps. Hence, it follows that

$$t_\pi(a_1, \ldots, a_k) \leq 1 + \sum_{i=1}^{m'} f_{n-1}^{(i(q+1))}(m') + m' + 1,$$

where $m' = \max(a_1, \ldots, a_{k-1}, a_k)$. Using properties of the $f_n^{(p)}(q)$, we get

$$t_\pi(a_1, \ldots, a_k) \leq f_n^{(p)}(m')$$

for some appropriate p (which can easily be determined). To finish the proof, we still have to consider composition of programs. So let $\pi_1, \pi_2 \in L_n$ and consider the program π defined by

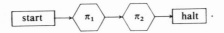

The program π is again an element of L_n. We obtain a bound on t_π as follows: Let $y_1, \ldots, y_r$ be the input variables of π_2 and $x_1, \ldots, x_k$ be the

input variables of π_1. Consider the r functions $g_{\pi_1}^{(1)}, \ldots, g_{\pi_1}^{(r)}$ computed by π_1 with output variables $y_1, \ldots, y_r$, respectively. Then

$$t_\pi(a_1, \ldots, a_k) = t_{\pi_1}(a_1, \ldots, a_k) + t_{\pi_2}(g_{\pi_1}^{(1)}(a_1, \ldots, a_k), \ldots, g_{\pi_1}^{(r)}(a_1, \ldots, a_k))$$
$$< f_n^{(p)}(\max(a_1, \ldots, a_k)) + f_n^{(q)}(f_n^{(r)}(\max(a_1, \ldots, a_k)))$$
$$\text{for some} \quad p, q, r$$
$$< f_n^{(s)}(\max(a_1, \ldots, a_k)) \quad \text{for some} \quad s.$$

6.5 Theorem. The function g_π computed by a loop program π of loop complexity n is majorized by the function f_{n+1}.

Proof. It is sufficient to show that f_{n+1} majorizes every function $f_n^{(p)}$. For $n = 0$ this is clear; namely, recall that $f_1(x) = 2x$ for $x \geq 2$ and $f_0^{(p)}(x) = x + 2p$. Thus f_1 majorizes $f_0^{(p)}$ for any fixed p. For $n > 0$, we proceed by induction on p. First, for $p = 0$, we have $f_n^{(0)}(x) = x < f_{n+1}(x)$ by (4). Next, assume that f_{n+1} majorizes $f_n^{(p)}$. Now observe that

$$f_n^{(p+1)}(x) < f_n^{(p+1)}(2 \cdot (x-2)) \quad \text{for} \quad x \geq 5$$
$$= f_n^{(p+1)}(f_1(x-2)) \leq f_n^{(p+1)}(f_n(x-2)) = f_n^{(2)}(f_n^{(p)}(x-2))$$
$$< f_n^{(2)}(f_{n+1}(x-2)) \quad \text{by induction hypothesis}$$
$$= f_n^{(2)}(f_n^{(x-2)}(1)) = f_n^{(x)}(1) = f_{n+1}(x) \quad \text{by definition.}$$

6.6 Theorem. Each function f_n can be computed by a loop program of loop complexity n.

Proof. The reader is invited to provide a program of loop complexity 1 for $f_1(x) = 2x + (1 \dotdiv x)$. Suppose that π_n is a program for f_n; we compose a program for f_{n+1} as follows:

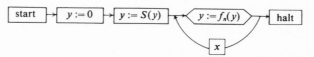

This program obviously computes $y := f_n^{(x)}(S(0)) = f_n^{(x)}(1) = f_{n+1}(x)$.

From Theorem 6.6, it follows that each function f_n is a primitive recursive function. Let us now consider the diagonal function d defined by

$$d(n) = f_n(n).$$

Is d a primitive recursive function? If it were, then there would exist n and p such that

$$d(x) < f_n^{(p)}(x) \qquad \text{for all} \quad x.$$

Now let N be any number greater than $n+1$ such that

$$f_{n+1}(x) > f_n^{(p)}(x)$$

for all $x \geqslant N$. Such N exist by the proof of Theorem 6.5. Now consider the following inequalities:

$$d(N) = f_N(N) < f_n^{(p)}(N) < f_{n+1}(N).$$

But if $N > n+1$, then $f_N(x) > f_{n+1}(x)$ as easily verified, and we have a contradiction. Thus, the function d is not primitive recursive; it grows much too fast for one. However, the function d is easily shown to be computable [see (c) of Problem 6.8]. Thus we have

6.7 Theorem. There exists a function which is computable but not primitive recursive.

6.8 Problems
(a) Prove observations (5) and (7) above.
(b) Write a loop program (of loop complexity 1) for the function $f_1(x) = 2 \cdot x + (1 \dotdiv x)$.
(c) Write a program for the diagonal function d.

7. STORED PROGRAMS

In the present section we wish to take advantage of the fact that computer programs are formal objects, strings of symbols as it were, and are therefore themselves capable of being stored and operated on by the computer. The particular model of a computer with which we are dealing in this chapter has, for reasons of simplicity, no capability to store or deal with strings of letters; it is a purely numerical computer. This is no drawback in principle because we can easily encode strings of letters as natural numbers and then work with the code numbers of programs rather than with the programs themselves. Such a process of encoding formal expressions into natural numbers is generally called Gödel numbering (after that giant of mathematical logic Kurt Gödel who first employed this technique in 1931 to obtain one of the deepest theorems of twentieth century mathematics).

Below, we shall first represent the details of a Gödel numbering for programs. This numbering is chosen in such a fashion that coding (i.e., translating programs into natural numbers) and decoding (i.e., translating from numbers back to programs) are easily accomplished. Of course, many different such Gödel numberings can be invented. We here present one that depends on the unique representability of natural numbers by the product of its prime factors.

The eventual aim of this section is to obtain a program ω that accomplishes the following: ω acts on essentially two memory locations, x and y. At x is stored the Gödel number of some program π, and at y is stored the Gödel number of the input $\langle a_1,...,a_m \rangle$ of π. Using some additional auxiliary memory cells, the program ω mimics the action of π on the input by changing the content of x for each step of π from the Gödel number of the input to this step to the Gödel number of the output of the step. Our task now is to spell out the exact way in which the outline above can be realized.

(1) Let p_n be the function which gives us the nth prime. Thus $p_1 = 2, p_2 = 3, p_3 = 5, ...$

(2) Consider now any program π on, say, the first m variables $x_1, ..., x_m$. At each instant of computation, these variables are assigned some values, $a_1, ..., a_m$. We encode this finite sequence of numbers by $2^{a_1} \cdot 3^{a_2} \cdot \cdots \cdot p_k^{a_k} \cdot \cdots \cdot p_m^{a_m}$. In this fashion the complete memory content is given by a single number.

(3) Now let us see how a single operating instruction of π would change the encoded memory content.

7. STORED PROGRAMS

(a) If the instruction is $x_k := 0$, then its execution changes $2^{a_1} \cdots p_k^{a_k} \cdots p_m^{a_m}$ to $2^{a_1} \cdots p_k^0 \cdots p_m^{a_m}$, that is, to the result of repeatedly dividing out by p_k until p_k is no longer a factor.

(b) If the instruction is $x_k := S(x_k)$, then its execution is mimicked by multiplication with p_k, if it is $x_k := P(x)$, then by division by p_k if p_k is a factor.

(c) If the instruction is $x_i := x_j$, then $2^{a_1} \cdots p_m^{a_m}$ would first be divided out by the highest power of p_i and afterward multiplied by as many copies of p_i as there are factors p_j.

(4) Next we encode the individual instructions by their Gödel numbers as shown in Table 2.1.

TABLE 2.1

Instruction	Gödel number
go to p	2^p
do $x_k := 0$ then go to p	$2^p \cdot 3^{k+1} \cdot 5$
do $x_k := S(x_k)$ then go to p	$2^p \cdot 3^{k+1} \cdot 7$
do $x_k := P(x_k)$ then go to p	$2^p \cdot 3^{k+1} \cdot 11$
do $x_i := x_j$ then go to p	$2^p \cdot 3^{i+1} \cdot 13^{j+1}$
if $x_k = 0$ then go to p else go to q	$17^{k+1} \cdot 19^p \cdot 23^q$
halt	29

(5) Any given instruction and memory content determines uniquely the label of the next instruction and a new memory content. If these entities are represented by the Gödel numbers, the interpretation of the action of instructions may be considered as numerical functions

$$y := \text{exec}_1(b, y)$$

where b is the Gödel number of the instruction, y the encoded memory content, and

$$a := \text{exec}_2(b, y)$$

where a is the label of the next instruction. The functions exec_1 and exec_2 are well determined by our conventions. Note that $\text{exec}_2(b, y)$ computes the exponent of 2 in b if 2 is a factor. Otherwise it first finds the exponent of 17, say $k+1$, seeks the exponent of p_k in y, and checks whether it is zero or not; in the first case the value of $\text{exec}_2(b,y)$ is the exponent of 19 in b, in the second case it is that of 23. If neither 2 nor 17 is a factor, then $\text{exec}_2(b,y)$ is set to zero. The function exec_1 is equally easy to describe: If b is not divisible by $6 = 2 \cdot 3$, then $\text{exec}_1(b,y)$ is simply y. In case it is divisible by 2 and 3, we first find the exponent of 3, say $k+1$, and try which of 5, 7, 11, or 13 also divides b. The value of $\text{exec}_1(b,y)$ is then computed accordingly; see item (3).

(6) Let us now assume that we are given a program π and that the labels of instructions in π are in the set $\{1, 2, \ldots, n\}$. Let b_i be the Gödel number according to (4) of the instruction with label i, and let <u>start</u>: <u>go to</u> p be the start instruction of π. Then the Gödel number $\bar{\pi}$ of π is defined as

$$\bar{\pi} = 2^{2^p} \cdot 3^{b_1} \cdot 5^{b_2} \cdots p_{k+1}^{b_k} \cdots p_{n+1}^{b_n}.$$

We shall need the function $\text{instr}(\bar{\pi}, k)$ which, for every program (with Gödel number $\bar{\pi}$) and label k, gives us the (Gödel number of the) instruction with label k. It is simply the exponent of p_{k+1} in $\bar{\pi}$.

The definitions and observations (1)–(6) above can now be brought together and used to formulate the program ω. In this program we use the variables x, y, u, and v as follows:

x stores the Gödel number $\bar{\pi}$ of the program to be executed,
y stores the encoded memory content $2^{a_1} \cdot 3^{a_2} \cdots p_m^{a_m}$,
u stores the label k of the instruction to be executed,
v stores the Gödel number b of that instruction.

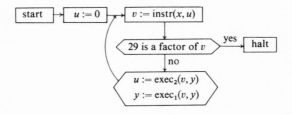

Fig. 2.57

7. STORED PROGRAMS

In terms of the capabilities instr, exec_1, and exec_2 discussed above, the program ω is composed as shown in Fig. 2.57.

Before we look into the details of the subroutines instr, exec, and so forth, let us discuss the overall behavior of this program ω.

Let π be any program with m variables, and let $\langle a_1,\ldots,a_m\rangle$ be the input to π. Now consider the program ω. Its input consists of $\bar{\pi}$ as input value for x and $2^{a_1}\cdots p_m^{a_m}$ as input value for y. In a first step u is set to zero and the loop of ω started.

The first step in the loop consists in computing $\text{instr}(x,u)$, that is, the exponent of p_{k+1} in the prime factor representation of x where k is the present value of u, namely 0. This gives us the value 2^r for some r (namely, the label of the first actual instruction of π to be executed).

Now, 29 is not a factor of $v = 2^r$ and thus we compute $y := \text{exec}_1(2^r, y)$. Since 2^r is not divisible by 6, y remains the same, namely $2^{a_1}\cdots p_m^{a_m}$.

Next we compute $u := \text{exec}_2(2^r, 2^{a_1}\cdots p_m^{a_m})$. The value is p, as it should be, the label of the next instruction.

Let us assume that r is the label of an instruction

$$r: \underline{\text{do}}\ x_k := P(x_k)\ \underline{\text{then go to}}\ q.$$

Thus $\bar{\pi}$ has a factor of the form $p_{r+1}^{2^q \cdot 3^{k+1} \cdot 7}$.

If we now start our second run through the loop of ω and compute $\text{instr}(\bar{\pi}, r)$, we get a new value for v, namely $2^q \cdot 3^{k+1} \cdot 7$. Again, 29 is not a factor of v, and we can proceed to evaluate exec_1 and exec_2. Now, $\text{exec}_1(2^q \cdot 3^{k+1} \cdot 7,\ 2^{a_1}\cdots p_m^{a_m})$ is $2^{a_1}\cdots p_m^{a_m}$ divided by p_k. This value, $2^{a_1}\cdots p_k^{(a_k-1)}\cdots p_m^{a_m}$, is the encoding of the proper value for the memory cells after the instruction in question has been performed. Next, $\text{exec}_2(2^q \cdot 3^{k+1} \cdot 7,\ 2^{a_1}\cdots p_m^{a_m})$ gives q, the label of the next instruction to be followed.

In case that r is the label of an instruction

$$r: \underline{\text{if}}\ x_k = 0\ \underline{\text{then go to}}\ q_1\ \underline{\text{else go to}}\ q_2,$$

then $\bar{\pi}$ would have a factor of the form $p_{r+1}^{17^{k+1} \cdot 19^{q_1} \cdot 23^{q_2}}$, and the run through the loop would be as follows.

First $\text{instr}(\bar{\pi},r)$ computes a new value for v, the exponent of $r+1$ in $\bar{\pi}$, thus $v = 17^{k+1} \cdot 19^{q_1} \cdot 23^{q_2}$. Of course 29 does not divide v and we compute exec_1. Since 6 is not a factor either, $\text{exec}_1(v, 2^{a_1}\cdots p_m^{a_m})$ gives $2^{a_1}\cdots p_m^{a_m}$

again. As for $\exec_2(17^{k+1} \cdot 19^{a_1} \cdot 23^{a_2}, 2^{a_1} \cdots p_m^{a_m})$, we obtain the value q_1 if a_k is zero, q_2 if it is not. With these values we now enter the loop again.

If the instruction with label r is of the form

$$r: \underline{\text{halt}},$$

then $\bar{\pi}$ has a factor of the form p_{r+1}^{29}. The first step would compute $v = 29$ and in the next step the decision would be to halt the program ω. And indeed, at the corresponding step the program π would also halt.

The other types of instructions are treated in exactly the same fashion. Thus, if we can come up with programs for the subroutines that enter the program ω, we have proved the following.

7.1 Theorem. There exists a program ω which is universal in the following sense. Suppose that π is a program with variables $x_1, \ldots, x_m$ and that π terminates on input $\langle a_1, \ldots, a_m \rangle$ with output $\langle b_1, \ldots, b_m \rangle$. Let $\bar{\pi}$ be the Gödel number of π. Then ω terminates on input $\langle \bar{\pi}, 2^{a_1} \cdots p_m^{a_m} \rangle$ with output $\langle \bar{\pi}, 2^{b_1} \cdots p_m^{b_m} \rangle$; if π does not terminate, then neither does ω.

It remains to find programs for the subroutines

$$\rightarrow \boxed{y := \text{instr}(x, u)} \rightarrow$$

$$\rightarrow \boxed{y := \exec_1(v, y)} \rightarrow$$

$$\rightarrow \boxed{u := \exec_2(v, y)} \rightarrow$$

and for the branching

$$\boxed{29 \text{ is a factor of } v} \xrightarrow{\text{yes}}$$
$$\downarrow \text{no}$$

This last subroutine can easily be composed out of the known loop program [see (d) of Problem 2.11] for the primitive recursive function rm_p for $p = 29$. The details for the remaining functions are as follows:

(a) The function p_n giving the nth prime is primitive recursive, and

(b) The function $n = [x, y]$ giving the greatest number n such that y^n divides x is primitive recursive.

By Theorem 5.2 it is sufficient to provide loop programs for these functions. Some functions such as addition, subtraction, multiplication, sg, $\overline{sg}$, ε, $\overline{\varepsilon}$, and so forth, are already known to us as primitive recursive; we may use corresponding subroutines to make the resulting programs more perspicuous. The factorial $y := x!$, being defined by $f(0) = 1$, $f(S(x)) = (x + 1) \cdot f(x)$, is also primitive recursive.

As a first step we exhibit an auxiliary program which detects whether a given number is prime, in the following sense: The function $prime\,(x)$ has value 1 if x is a prime, 0 if it is not. The loop program shown in Fig. 2.58 computes $y := prime(x)$. This program first checks whether

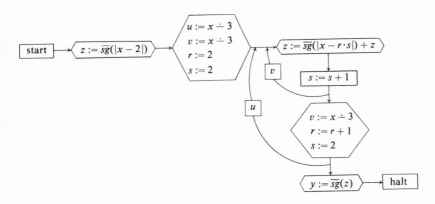

Fig. 2.58

$x = 2$. If it is, then z is set to 1 or else to 0. After this, the program searches for factors r and s of x. As long as none has been found, the value of z remains at zero, otherwise this value becomes (and remains) positive. Thus, at the end $\overline{sg}(z)$ is 1 exactly if no nontrivial factors of x have been found.

The program given in Fig. 2.58 can be used as a subroutine to compute the function p_n which gives us the nth prime, $z := p_n$, as shown in Fig. 2.59. The inner loop computes the next prime p_{n+1} given $z = p_n$. For this purpose it searches $t = z+1, z+2, z+3, \ldots, z+z!$ for a prime. If a prime t is found, that is, if $u = \text{prime}(t) = 1$, the instruction $z := sg(u) \cdot t + \overline{sg}(u) \cdot z$ sets $z := t$, which is the value of p_{n+1}. Until then the value of z is left unchanged. After t has been found as prime the

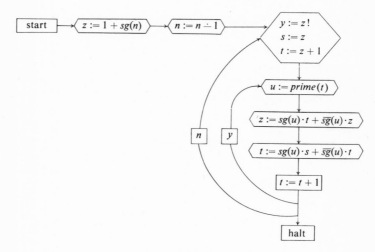

Fig. 2.59

instruction $t := sg(u) \cdot s + \overline{sg}(u) \cdot t$ sets the value of t back to s, that is, back to p_n. After $p_n!$ steps the next prime must be found. For suppose that none of the numbers $p_n + 1, p_n + 2, \ldots, p_n! + 1$ are prime. Then $p_n! + 1$ must be divisible by a prime $k \leq p_n$. But notice that

$$\frac{p_n! + 1}{k} = \frac{2 \cdot 3 \cdots (k-1) \cdot k \cdot (k+1) \cdots p_n + 1}{k}$$

$$= 2 \cdot 3 \cdots (k-1)(k+1) \cdots p_n + \frac{1}{k}$$

is not an integer, a contradiction.

Finally, we present a loop program (Fig. 2.60) for the function $n := [x, y]$ which gives us the largest integer n such that y^n divides x.

For the proof of the theorem above, we need not have insisted, as we did repeatedly, that the functions entering ω, namely, instr, $\exec_1$, and $\exec_2$, are primitive recursive. This observation, however, is used below in the formulation of a refinement of Theorem 7.1, namely, Kleene's normal form theorem below.

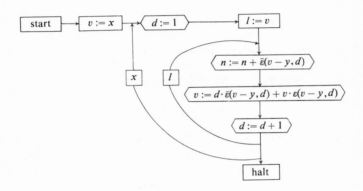

Fig. 2.60

Let us recall the universal program ω. For any program π and input $\langle a_1, \ldots, a_m \rangle$ to π, it mimics the computation steps that π goes through and thus, in the given encoding, computes the same function as π does. We shall now take a closer look at how this computation by ω proceeds. Suppose therefore that π and $\langle a_1, \ldots, a_m \rangle$ are given. In a first step we compute the Gödel number

$$g(a_1, \ldots, a_m) = 2^{a_1} \cdots p_m^{a_m}$$

of the input. This is the input value of y for ω; the input value of x for ω is $\bar{\pi}$, the Gödel number of the program π. With the input thus computed, we start the program ω as shown in Fig. 2.61. The value of y upon entering

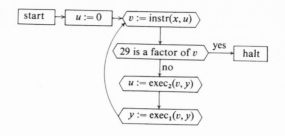

Fig. 2.61

the loop for the first time is $y_0 = g(a_1,\ldots,a_m)$; the value of v at this time is $v_0 = \text{instr}(\bar{\pi},0)$. Each time the loop is passed through, the values of y and v are computed anew. We thus obtain a sequence

$$\langle y_0, v_0 \rangle, \langle y, v_1 \rangle, \ldots$$

This sequence, if it terminates at all, terminates with the first v_j which is divisible by 29. We denote the function which gives us the value v_j by $h_1(j,a_1,\ldots,a_m,\bar{\pi})$. Let $rm_p(x)$ denote the remainder of x after division by p. The number of terms in the sequence is therefore

$$(\mu j)[rm_{29}(h_1(j,a_1,\ldots,a_m,\bar{\pi})) = 0].$$

Let us denote by $h_2(j,a_1,\ldots,a_m,\bar{\pi})$ the function that gives us the value of y_j. Then the value of y at termination is given by

$$h_2((\mu j)[rm_{29}(h_1(j,a_1,\ldots,a_m,\bar{\pi})) = 0], a_1,\ldots,a_m,\bar{\pi}).$$

In case we are only interested in the output value at x_0 for π, then this value would be

$$e_2(h_2((\mu j)[rm_{29}(h_1(j,a_1,\ldots,a_m,\bar{\pi})) = 0], a_1,\ldots,a_m,\bar{\pi})).$$

Now the function $e_2(n)$ (exponent of 2 in the prime factorization of n) and rm_{29} are primitive recursive. We show below that h_1 and h_2 are also primitive recursive. Hence we have the following result.

7.2 Kleene's Normal Form Theorem. For every $m > 0$, there exists primitive recursive functions h_1 and h_2 such that, for every partial recursive function f, there is a number e with

$$f(a_1, \ldots, a_m) = h_2((\mu j)[h_1(j, a_1, \ldots, a_m, e) = 0], a_1, \ldots, a_m, e)$$

for all $a_1, \ldots, a_m$ on which f is defined (if f is not defined on $a_1, \ldots, a_m$, then neither is the right-hand side of the equation).

(Thus, the μ-operator—which is not applied to obtain primitive recursive functions—need only be applied once to obtain any partial recursive function whatsoever.)

To finish the proof, we need to verify that the functions h_1 and h_2 are indeed primitive recursive. We see at once that they are defined by simultaneous recursion, namely

$$h_1(0, a_1, \ldots, a_m, \bar{\pi}) = \text{instr}(\bar{\pi}, 0)$$
$$h_2(0, a_1, \ldots, a_m, \bar{\pi}) = g(a_1, \ldots, a_m)$$
$$h_1(S(n), a_1, \ldots, a_m, \bar{\pi}) = \text{instr}(\bar{\pi}, \text{exec}_2(h_1(n, a_1, \ldots, a_m, \bar{\pi})))$$
$$h_2(S(n), a_1, \ldots, a_m), \bar{\pi}) = \text{exec}_1(h_1(n, a_1, \ldots, a_m, \bar{\pi}), h_2(n, a_1, \ldots, a_m, \bar{\pi})).$$

We have already verified that the functions instr, exec_1, and exec_2 are primitive recursive.

To finish this section, let us make some timely remarks on the encoding of programs that we used. Clearly, Gödel numbers of even very short programs get to be very large, unmanageably so. For example, the Gödel number of the following program

start: go to 1;
1: do $x_1 := P(x_1)$ then go to 2;
2: halt.

is $2^{2^1} \cdot 3^{2^2 \cdot 3^2 \cdot 11} \cdot 5^{29}$, which is a number with more than 200 digits. Much of the information content of these 200 digits is, of course, irrelevant, and a more proper encoding is clearly called for. Thus, while ω does, perhaps, remind us conceptually of actual stored program machines, there is very little that it has to do with the actual storage and retrieval of information and commands in present-day computers.

7.3 Problem. Prove Theorem 2.12, page 114, using Theorem 7.2.

8. THE THESIS OF CHURCH AND TURING

Let P be a property of programs, such as "π computes $\sqrt{x}$," "π has three exits," "π has no loops." To each program π, we associate its Gödel number $\bar{\pi}$. Thereby, there corresponds to P a set Q of numbers, namely the set of Gödel numbers of programs π which have the property P. Assume now that we are interested in deciding whether a given program π has property P. In terms of Gödel numbers, this amounts to deciding whether $\bar{\pi}$ is in Q. One type of method comes to mind, which we formulate as a definition.

8.1 Definition. A set Q of numbers is said to be recursively decidable (or simply "recursive") if there exists a program δ with two exits, "yes" and "no," such that for any input n the program δ halts in either one or the other exit; it halts in "yes" if n is in Q, in "no" if it is not. A set Q is said to be recursively undecidable if no such program exists.

The notions of recursive decidability and undecidability are transferred to properties of programs via Gödel numbering. Thus we say that a property P is recursively decidable (undecidable) if and only if its corresponding set of Gödel numbers is recursively decidable (undecidable).

There is a possible source of ambiguity in this convention which we have to clear up first. So far we have discussed only one Gödel numbering of programs, the one in the previous section. One does not need much imagination to observe that there are many other and quite different kinds of schemes whereby we may assign a natural number to a program. Is it therefore not conceivable that a property P is recursively undecidable according to one Gödel numbering, and decidable according to another? Clearly yes, if we understand the term Gödel numbering broad enough, broader than is practical. For, what do we want a Gödel numbering to do: We want it to provide an easily accomplished translation from π to $\bar{\pi}$. If we have a second Gödel numbering, from π to $\bar{\bar{\pi}}$ say, then we have in fact defined a one-to-one function t associating $\bar{\pi}$ with $\bar{\bar{\pi}}$. It is not too much to ask that t itself be computable by a program. (More on this later.) But with the aid of a program for t any decision program for P under the Gödel numbering $\bar{\bar{\pi}}$ can be used to obtain a decision program for P under the Gödel numberings for which the translations to and from our fixed Gödel numbering are computable. In this sense, we are justified in talking about the recursive decidability or undecidability of properties without mentioning the Gödel numbering.

Let us now consider one particular and quite central property of programs, namely, its termination behavior. Suppose we are given a program π and an input $\langle a_1,\ldots,a_m \rangle$ for π. Will π halt on this input? A superficial answer is quickly given: Why not simply start π on $\langle a_1,\ldots,a_n \rangle$ and observe whether it halts or not? Well, if it halts, we know the answer; but if it has not halted by December 31, 1999, do we know for sure that it will not halt on January 1, 2000? So clearly this is not the right approach. What we would like to have is a decision program δ as in Definition 8.1 which would give us a uniform method to answer (and always give answer) the question at hand.

8.2 Theorem. The halting problem for programs, that is, the problem as to whether a given program with given input ever halts, is recursively undecidable.

Proof. Let us call a decision program any program δ with exactly two exits, "yes" and "no."

8.3 Lemma. There is no decision program δ which accomplishes the following: Suppose that π is a decision program. Let us start δ with input $\bar{\pi}$ for all variables. Then δ goes into exit "yes" iff π terminates in exit "no" for this same input.

The lemma is easily proved. Assume, namely, that δ were this program and let it be started on input $\bar{\delta}$. According to its definition δ would go into exit "yes" on input $\bar{\delta}$ iff δ terminates in "no" for this input. This is absurd, and therefore no such δ can exist.

Suppose we had a decision program γ for the halting problem, that is, a program which terminates on input $\langle a,b \rangle$ iff a is the Gödel number of a program which terminates on the input with Gödel number b. We may assume without loss of generality that the output of γ is again $\langle a,b \rangle$.

Let us now recall the universal program ω. It has only one exit; thus it is, superficially, unable to mimic the decision behavior of a decision program. However, it needs only very little alteration to adapt it to this purpose. For suppose that π is a decision program with two exits

$$r: \underline{\text{halt}};$$
$$s: \underline{\text{halt}};$$

where the first one is the "yes" exit and the second one the "no" exit. Let us change the Gödel number of π from

$$2^{2^p} \cdots p_r^{29} \cdots p_s^{29} \cdots p_n^{b_n}$$

to

$$2^{2^p} \cdots p_r^{29^2} \cdots p_s^{29} \cdots p_n^{b_n},$$

and let us revamp ω into the program shown in Fig. 2.62. Thus, clearly, ω goes into exit "yes" ("no") on input $\langle \bar{\pi}, 2^{a_1} \cdots p_m^{a_m} \rangle$ iff π goes into exit "yes" ("no") on input $\langle a_1, \ldots, a_m \rangle$.

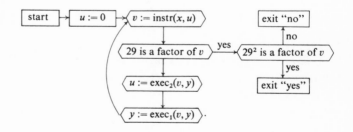

Fig. 2.62

8. THE THESIS OF CHURCH AND TURING

Let $f(m) = 2^m \cdots p_n^m$ if m is the Gödel number of a program π and π contains variables $x_1, \ldots, x_n$; $f(m) = 0$ if m is not a Gödel number of a program. Now let us compose our hypothetical program γ (variables x, y) with the modified universal program ω (variables $x, y, u, v, \ldots$) as shown in Fig. 2.63. Suppose we start this program δ with the input $\bar{\pi}$ for all variables. Now, if π does not terminate on this input, then the program δ goes into exit "no." If π does terminate, we use ω to determine which exit π takes. If ω takes exit "yes," that is, if π takes exit "yes," then we let δ take exit "no"; if ω takes exit "no," that is, if π takes exit "no," then we let δ take exit "yes." Thus δ goes into exit "yes" on input $\langle \bar{\pi}, \bar{\pi} \rangle$ iff π goes into exit "no" on input $\bar{\pi}$.

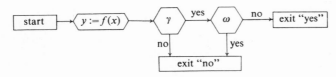

Fig. 2.63

Our composite program δ would therefore be a decision program whose existence was denied in the lemma. Hence our hypothetical program γ cannot exist either, and we are through.

Church's thesis, which is formulated below, suggests that we drop the qualification "recursively" from our statement of the theorem. We are thus invited to state, without any qualifications or reservations: "There is none and never will be any decision procedure for the halting problem for programs." This statement belongs to a realm of transmathematical pronouncements of which we may or may not be believers. By its very form it cannot make up the content of a provable or disprovable statement in any present-day field of mathematics: The expression "decision procedure," while it has a quite definite intuitive connotation, lacks a firm codification within a mathematical framework. The best we can therefore do is to adduce evidence which shows that by whatever means we have tried to codify the idea of decision procedure we always wind up with a notion that is an inessential variant to that of a recursive decision procedure.

Before we embark on such a course, let us make sure as to how, exactly, we should define our goal. To begin with, we may consider the making of a decision on whether a number n belongs to a set Q as the computation of a function c: $c(n) = 0$ if n is in Q, $c(n) = 1$ otherwise. Thus the question as to what constitutes a decision procedure becomes absorbed in the question as to what a computable function is. And it is in terms of this question that Church's thesis is formulated.

8.4 Church's Thesis.[6] A numerical partial function is computable iff it is partial recursive.

(Originally Church stated this thesis for total functions only; it is extended to partial functions in the following sense: The precept that governs the computation procedure leads to no value for the function iff as a partial recursive function it is undefined for that argument.)

We have seen earlier that the notions of partial recursive function and function computable by a program on the universal calculator are equivalent. This may serve as a first piece of evidence of the sort that we would admit in favor of Church's thesis. Let us now bring forth another piece of evidence of the same nature, namely the fact that the clarification of the computability concept by means of Turing machines does not lead outside of the realm of partial recursive functions either.

Turing arrived at the notion of a Turing machine by way of an analysis of the intuitive concept of the processing of data.[7] Let us briefly recount the basic steps in this analysis.

Imagine a being or machine called "processor" that is able to act according to a fixed plan on a set of data, eventually producing a transformation of these data. Let us analyze the notion of data and transformation of data into their atomic, that is, indivisible, constituents. The smallest amount of data that a processor can recognize at any one instant is called a symbol. It is reasonable to posit that there are only finitely many symbols. The set of symbols that constitute the data is arranged in some manner to form the data. This manner cannot be

---

[6] The reader may wish to read some of the major original papers on the general subject of computability and decidability in the collection of Davis [2].

[7] See Turing's original paper [15], reprinted in [2].

quite arbitrary since the processor must not lose his orientation among the data. The more complicated this arrangement in space is, the more involved will be the type of plan according to which the processor's attention moves around. Let us be bold and assume that all data are arranged in a linear array. That is, imagine a tape, subdivided into cells numbered, ..., $-3, -2, -1, 0, 1, 2, 3, \ldots$. Each cell contains only one symbol or is blank. It is reasonable to assume that the amount of data is finite, hence all but finitely many cells will be blank. Now, to analyze the notion of processing such a data structure into its atomic acts, three actions suggest themselves. The processor's attention at any one moment is fixed on exactly one cell (since the maximal amount of data that the processor can recognize at any one moment is what a single cell may contain). Thus the processor may either change the symbol in this cell or move his attention to the cell immediately to the left or to the cell immediately to the right of the cell under scan. (The move to any other cell and changing its content is clearly a composite action of possibly great complexity.) Thus our assumption that data are arranged linearly leads to considerable economy in the number of atomic actions. (The reader is invited to experiment with some arrangements of data of his own invention and their corresponding atomic acts. He will convince himself that such variants reduce by proper encoding to the arrangement and actions above; the reduction may ask for some ingenuity on the part of the inventor.)

We have now arrived at the constituents of the definition of a Turing machine, except for one thing: How does the processor "know" which of the actions he is supposed to take at any one instant? The determination to perform a particular action is a "state of mind," for the human being and, by extension, of any processor. It is reasonable to assume—minds being finite—that there are only finitely many states possible to the processor. A plan for processing given data therefore will have to be formulated in terms of assigning actions and changes of state to all combinations of present states and observed data.

8.5 Definition. (i) A *Turing machine* $\mathcal{T}$ over an alphabet $A = \{a_0, \ldots, a_m\}$, where a_0 is the symbol standing for the blank, consists of

(a) S, a finite set of *states*, $s_0 \in S$ the *start* state, $T \subseteq S$ the set of halting states;

(b) a next-state function σ which maps $S \times A$ into S;

(c) a next-action function μ which maps $S \times A$ into $A \cup \{r,l\}$, where r and l are assumed different and do not belong to A.

(ii) A *tape* for a Turing machine over A is a map t from the set of integers into A such that $t(i) = a_0$ for all but finitely many integers i.

(iii) A *configuration* of a Turing machine is a triplet $\langle s, t, i \rangle$ consisting of a state s, a tape t, and an integer i. The symbol $t(i)$ is called the *scanned symbol* of this configuration, and s is called the present state of the Turing machine.

(iv) A *computation* by a Turing machine $\mathcal{T}$ is a finite or infinite sequence of configurations $\langle s_0, t_0, i_0 \rangle, \langle s_1, t_1, i_1 \rangle, \ldots$ satisfying the following requirements:

(a) s_0 is the start state of $\mathcal{T}$; the pair $\langle t_0, i_0 \rangle$ is called the *input of* the computation.

(b) s_{k+1}, t_{k+1}, and i_{k+1} are determined by s_k, t_k, and i_k using the functions σ and μ. Thus,

$$s_{k+1} = \sigma(s_k, t_k(i_k)).$$

(The next state is determined by the present state and the scanned symbol.)

$$t_{k+1} = t_k \quad \text{if} \quad \mu(s_k, t_k(i_k)) = r \quad \text{or} \quad l;$$

if $\mu(s_k, t_k(i_k)) = a_s$, then t_{k+1} is the same as t_k except that the value at i_k is changed to a_s.

(The value of μ for the present state and scanned symbol gives the symbol a_s with which the scanned symbol is to be replaced—if it is of the form a_s; in case it is not, then the tape is not changed.)

$$i_{k+1} = i_k \quad \text{if} \quad \mu(s_k, t_k(i_k)) \in A,$$
$$i_{k+1} = i_k + 1 \quad \text{if} \quad \mu(s_k, t_k(i_k)) = r,$$
$$i_{k+1} = i_k - 1 \quad \text{if} \quad \mu(s_k, t_k(i_k)) = l.$$

(If the value of μ is in A, then the action consists in changing the scanned symbol without moving attention to another place on the tape;

8. THE THESIS OF CHURCH AND TURING

in case it is r the next scanned symbol will be the one immediately to the right, if it is l, the one on the left of the presently scanned symbol.)

The computation is finite if there is a configuration of the form $\langle s_k, t_k, i_k \rangle$ with $s_k \in T$; it is said to *terminate* with the first such configuration and $\langle t_k, i_k \rangle$ is called the *output* of the computation. Otherwise the computation is infinite and is said not to terminate.

This rather long-winded definition renders the exact sense of the result of our analysis of the computation process. The remarks in parentheses are supposed to make the connection between the formal definition and its intuitive origin more apparent.

Taking the definition of Turing machines for granted, we now set ourselves the task to show that Turing machines can compute no more functions than the universal calculator can. The proof is very simple, once we understand the technique of Gödel numbering. Let us consider any configuration

$$\langle s, t, i \rangle$$

of a Turing machine. Only finitely many $t(j)$ are different from the blank symbol, and hence there are integers j_0 and j_1 such that all $t(j)$ with $j < j_0$ or $j > j_1$ are blank *and* such that $j_0 \leq i \leq j_1$. Let us choose the largest such j_0 and smallest such j_1. We now want to represent the total information given by t and i by a single number.

Let $\bar{a}_k = k + 1$ for all symbols $a_k \in A$. Consider the following auxiliary functions:

For $j \geq 0$ and $j \neq i$, let $g(t, i, j) = p_{2 \cdot j}^{2^{\overline{t(j)}}}$.

For $j \geq 0$ and $j = i$, let $g(t, i, j) = p_{2 \cdot j}^{2^{\overline{t(j)}} \cdot 5}$.

For $j < 0$ and $j \neq i$, let $g(t, i, j) = p_{-2 \cdot j + 1}^{2^{\overline{t(j)}}}$.

For $j < 0$ and $j = i$, let $g(t, i, j) = p_{-2 \cdot j + 1}^{2^{\overline{t(j)}} \cdot 7}$.

Finally, let

$$g(t, i) = \prod_{j_0 \leq j \leq j_1} g(t, i, j).$$

The number $g(t,i)$ encodes the wanted information in a very simple fashion. Suppose now that we are given a Turing machine $\mathcal{T}$ by its two functions, σ and μ, and let us try to mimic the behavior of $\mathcal{T}$ by a program for the universal calculator.

We reserve the variable x for the value of $g(t,i)$. The value of u will indicate the state of $\mathcal{T}$ and the value of v will indicate the action that $\mathcal{T}$ is to take. These values change during the mimicking computations as follows:

$$x := h_1(x,v)$$
$$u := h_2(x,u)$$
$$v := h_3(x,u).$$

The value of u is 2^i if $\mathcal{T}$ is in the ith state. The value of v is 2^{k+1} if the action is to change the scanned symbol to a_k; it is 3 if a move to the right is wanted, 5 if it is a move to the left. In terms of these conventions the functions h_1, h_2, h_3 are the following: $h_1(x, 2^{k+1})$ results from x by replacing the factor $p_j^{2^r \cdot 5}$ by $p_j^{2^{k+1} \cdot 5}$ and a factor $p_j^{2^s \cdot 7}$ by $p_j^{2^{k+1} \cdot 7}$. Exactly one of the two cases occurs if x is of the form $g(t,i)$. We leave it to the reader to formulate the definition of $h_1(x,3)$ and $h_1(x,5)$.

Let $u = 2^i$ and let k be that number for which there is a j with $p_j^{2^{k+1} \cdot 5}$ or $p_j^{2^{k+1} \cdot 7}$ which is a factor of x; such a k is uniquely determined. Then

$$h_2(x,u) = 2^{\sigma(s_i, a_k)}$$

and

$$h_3(x,u) = \begin{cases} 2^{p+1} & \text{if } \mu(s_i, a_k) = a_p, \\ 3 & \text{if } \mu(s_i, a_k) = r, \\ 5 & \text{if } \mu(s_i, a_k) = l. \end{cases}$$

It should be clear to the reader that we could, if pressed, write out actual programs for the operations h_1, h_2, and h_3. We now take these for granted and compose them as shown in Fig. 2.64. (A decision program for the question as to whether u is the number corresponding to a terminal state is easily arranged.)

Let the given Turing machine $\mathcal{T}$ be started on input t_0 with i_0 the scanned square. Simultaneously, let the program $\pi^{\mathcal{T}}$ be started with

8. THE THESIS OF CHURCH AND TURING

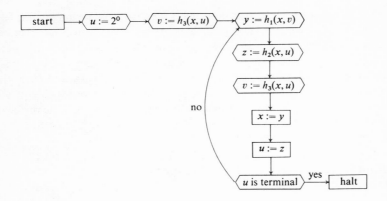

Fig. 2.64

input $x = g(t_0, i_0)$. When $\pi^{\mathcal{T}}$ reaches the branching point it will have executed the first action of $\mathcal{T}$ in the encoding given by g, and will have found numbers u and v representing the next state and next action of $\mathcal{T}$. If u represents a nonterminal state, this process of mimicking the behavior is repeated. Only if u represents a terminal state is the exit taken. In the corresponding computation of $\mathcal{T}$, meanwhile, this terminal state is also reached and $\mathcal{T}$ halts too. We have thus established the following.

8.6 Theorem. For every Turing machine $\mathcal{T}$ there exists a program $\pi^{\mathcal{T}}$ such that $\mathcal{T}$ halts on input $\langle t_0, i_0 \rangle$ with output $\langle t_k, i_k \rangle$ iff $\pi^{\mathcal{T}}$ halts on input $x = g(t_0, i_0)$ with output $x = g(t_k, i_k)$; if one of $\mathcal{T}$ and $\pi^{\mathcal{T}}$ does not halt, then neither does the other.

In view of our earlier remarks this result may be considered supporting evidence for Church's thesis: Whatever is computable by a Turing machine can also be computed by a program and hence is partial recursive. Our last statement is rather loosely worded; strictly speaking we have defined the notion of partial recursive function only for numerical functions while Turing machines essentially compute functions $f: A^* \to A^*$ for some finite alphabet A. Of course, with encoding of A^* by Gödel numbers, f turns into a numerical function. It is in this sense that we speak of f as a partial recursive function.

It is natural to ask whether, conversely, every partial recursive function is computable by some Turing machine. The answer is yes, but to prove this we would have to go through a considerable amount of detailed work to construct little Turing machines for diverse tasks, most of this work having no applications outside of the proof at hand. In any case, the reader can easily find this proof in the literature. In particular in the books of Davis [1], Hermes [3], and Kleene [4]. We provide one problem which will give the reader a taste of the kind of work that goes into designing Turing machines.

If we now take the alluded fact for granted, then Church's thesis is obviously equivalent to the following statement.

Turing's Thesis. A function $f: A^* \to A^*$ is computable iff it is computable by a Turing machine.

8.7 Problems

(a) Spell out the definition of $h_1(x, 3)$ and $h_1(x, 5)$.

(b) Consider Turing machines over the alphabet $\{0, 1, *\}$ where 0 stands for the blank. We give a tape t by writing a word such as

$$*100111*1\underline{0}1$$

indicating by an underline the scanned symbol. The understanding is that the tape to the left or right of the word above is blank. Devise a Turing machine which copies the word between asterisks to the left of the initially scanned symbol $*$ on the right of that spot. Thus, the word above would be transformed into

$$*100\ 111*100\ 111.$$

(c*) *The Bounded Halting Problem.*[8] Let us consider a program π and input a. One rarely asks: Does π *ever* halt on this input? Rather, one has in mind a time bound B and asks: Does π halt on a in at most B steps? This question, of course, is decidable: One only needs to run π on input a for B steps and observe whether it halts or not. But this test is of little

---

[8] Courtesy R. Jeroslow (unpublished).

practical value since it would cost as much computer time to test whether π is not too costly as it would to run π itself. What we should like to have, therefore, is a decision program as shown in Fig. 2.65 with the following

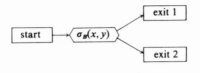

Fig. 2.65

properties: σ_B terminates in the upper exit on input $(\bar{\pi}, a)$ if π terminates on input a in at most B steps; if π takes more than B steps on input a, then σ_B exits through the lower exit. Such programs are quite easy to construct (do this by modifying ω) but they all suffer from the same flaw, which makes them quite useless.

Proposition. For every such program σ_B there exists a program π and an input a such that σ_B takes at least B steps on input $(\bar{\pi}, a)$.

Prove this proposition.

(d) By a *general recursive function* we understand a partial recursive function which is defined for all arguments. Prove that there is no computable function $f(x)$ with the following properties:

(i) $f(n)$ is the Gödel number of a program for a recursive function, for each n.
(ii) If $g(x)$ is a general recursive function, then there exists m such that $f(m)$ is the Gödel number of a program which computes g.

(e) Use the idea of the proof for problem (d) to give an alternative proof of the fact that there is a recursive function which is not primitive recursive.

(f) Construct an example of a numerical function which is not partial recursive.

(g) Find the error in the following argument: Let us define a function d by setting $d(n) = \omega(n, n) + 1$ where $\omega(x, y)$ is our universal program.

Since every partial recursive function $f(n)$ is $\omega(m,n)$ for some fixed m and $d(n)$ is obviously computable, we must have $d(n) = \omega(m,n)$ for some m. But clearly $d(m) \neq \omega(m,m)$, which is a contradiction.

(h) Show that the following definitions are equivalent.

(i) A set $M \neq \varnothing$ of natural numbers is *recursively enumerable* if $M = \{x: f(x) \text{ is defined}\}$ for some partial recursive f.

(ii) A set $M \neq \varnothing$ of natural numbers is *recursively enumerable* if $M = \{f(x): x \in N\}$ for some general recursive f.

9. RANDOM-ACCESS STORED PROGRAM MACHINES

We finally come to the problem touched upon at the end of Section 7, namely to describe a more realistic mathematical model of present-day computers. The added realism concerns the fact that these computers accept formal programs into their memories and operate by referring to the stored programs. While the program ω of recent fame had this capability, it had it in a very unrealistic fashion in that the storing and handling of the code $\bar{\pi}$ for the program π was exceptionally awkward. We want to remedy this.

There is an additional goal that we pursue. Namely, once the program (or a code for it) is stored in one or more memory cells of a computer there is nothing that prevents us, in principle, from letting the capabilities of the computer be applied to these particular memory cells also. This idea opens up the possibility that the computer itself changes, perhaps improves, the very program on which it operates. This suggests the exciting possibility that such self-modifying computers are more powerful tools than the fixed-program computers that we have dealt with so far. In a sense this is true. While such machines still compute nothing else but partial recursive functions, they do this, in general, more efficiently, that is, definitely in fewer computation steps.

In the present section, we present a model of machines which is designed to bring out the stored and self-modifying program feature most clearly and for which the results mentioned above will be proved.

9. RANDOM-ACCESS STORED PROGRAM MACHINES

Let us first give an intuitive description of this new type of machine.[9] A random-access stored program machine has a set of memory cells which may be filled with data or (encoded) instructions. When the machine is started, it first looks up the content of a particular memory cell, called the *instruction counter*, IC. The number n stored there then tells the machine to look up the content of the nth memory cell, Rn. The next step depends on what this cell contains. If it is an instruction calling for some changes in the stored data, the machine then performs that instruction; if it is the instruction calling for the termination of the computation, the machine *halts*; if it is not an instruction, the machine also halts, but in a different mode, called *jam*. If the machine does not halt or jam, it again looks up the present content of IC (which may be changed by the execution of the previous instruction) and repeats the procedure. Thus, for any given initial values to the memory cells, the machine will perform a series of computation steps which may either be infinite, or, in case it is finite, terminate in halt or jam.

The individual random-access stored program machine is given by specifying the set of instructions, their operation on the memory cells, and their encoding. We here present only one example of such a machine, which we call RASP.

The memory cells of RASP are divided into three groups,

the instruction counter	IC
the accumulator	AC
the memory registers	R1, R2, R2, ...

These cells contain words over the following alphabet

TRA, TRZ, STO, CLA, ADD, SUB, HALT, 0, 1, $\langle$, $\rangle$.

Words over the symbols 0, 1 are used to represent natural numbers in dyadic notation. The other symbols are used to form code words for instructions. Let $\bar{n}$ be the code for natural number n. We list below those instruction codes admitted for RASP together with their action on the memory cells. These instructions are such that IC will always contain a numeral.

---

[9] Abstract random-access stored program machines were first used by Elgot and Robinson [10]; the particular model used here is from Hartmanis [11].

TRA $\bar{n}$: This instruction sets the content of IC to $\bar{n}$.

TRA$\langle \bar{n} \rangle$: This instruction looks up the content of Rn and transfers its content to IC if it is a numeral, otherwise the machine is directed to jam.

TRZ $\bar{n}$: This instruction first looks up the content of AC; if that content is 0 then the content of IC is changed to $\bar{n}$; if the content of AC is different from 0 and if the content of IC is $\bar{m}$ for some m, then IC is changed to $\overline{m+1}$.

TRZ$\langle \bar{n} \rangle$: This instruction first looks up the content of AC; if that content is 0, then the content of Rn is looked up; if the content of Rn is a numeral $\bar{m}$, then IC is changed to $\bar{m}$, otherwise the machine is directed to jam; if the content of AC is different from 0 and IC contains $\bar{m}$, then the content of IC is changed to $\overline{m+1}$.

STO $\bar{n}$: This instruction copies the present content of AC into register Rn (destroying the previous content) and changes the content $\bar{m}$ of IC to $\overline{m+1}$.

STO$\langle \bar{n} \rangle$: This instruction first looks up the content of Rn; if this content is a numeral $\bar{m}$, then the present content of AC is copied into register Rm; if it is not a numeral, the machine is directed to jam.

CLA $\bar{n}$: This instruction changes the content of AC to $\bar{n}$ and the content of IC from $\bar{m}$ to $\overline{m+1}$.

CLA$\langle \bar{n} \rangle$: This instruction looks up the present content of register Rn and copies it into AC, then it changes the content $\bar{m}$ of IC to $\overline{m+1}$.

ADD $\bar{n}$: This instruction looks up the present content of AC; if that is a numeral $\bar{k}$, then the content of AC is changed to $\overline{k+n}$ and the content $\bar{m}$ of IC is changed to $\overline{m+1}$; if the content of AC is not a numeral, then the machine is directed to jam.

ADD$\langle \bar{n} \rangle$: This instruction looks up the contents of AC and Rn; if both are numerals $\bar{k}$ and $\bar{l}$, respectively, then AC is changed to $\overline{k+l}$ and IC is changed from $\bar{m}$ to $\overline{m+1}$; if one of these contents is not a numeral, then the machine is directed to jam.

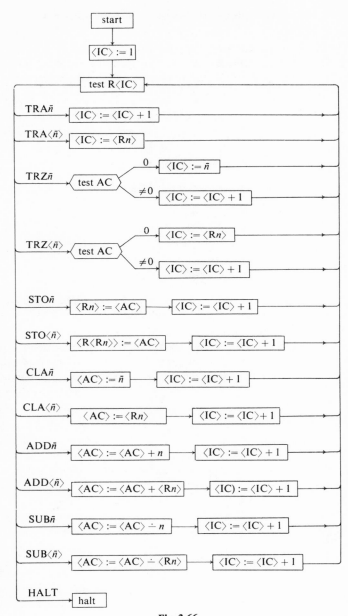

Fig. 2.66

SUB $\bar{n}$: This instruction looks up the present content of AC; if that is a numeral $\bar{k}$, then the content of AC is changed to $\overline{k \dotdiv n}$ and that of IC from $\bar{m}$ to $\overline{m+1}$; if the content of AC is not a numeral, then the machine is directed to jam.

SUB$\langle\bar{n}\rangle$: This instruction looks up the contents of AC and Rn; if both are numerals $\bar{k}$ and $\bar{l}$, respectively, then AC is changed to $\overline{k \dotdiv l}$ and IC is changed from $\bar{m}$ to $\overline{m+1}$; if one of these contents is not a numeral, then the machine is directed to jam.

HALT: This instruction directs the machine to halt.

This lengthy, but complete, description of the individual instructions and their action is contained in the flowchart shown in Fig. 2.66. In Fig. 2.66 we employ the notation $\langle Rn \rangle$, $\langle AC \rangle$, and $\langle IC \rangle$ for the *contents* of Rn, AC, and IC, respectively. The remainder is self-explanatory. (We have omitted all arrows to jam.)

The machine RASP is said to be *programmed* if we have assigned some instructions to the registers R1, ..., Rm. The sequence of contents of R1, ..., Rm is called the program of RASP.

The registers not used for the program are free to contain the initial data. As a rule, we specify some such registers (finitely many) to contain the input and call them *input registers*; all other registers are understood to contain zero. The n-tuple of contents of the input registers is called the *input* to the program. If we now start a programmed RASP on some input, the execution may either terminate in a halt or in a jam, or it may never terminate. In the first case, we consider the eventual content of AC as the output value of the computation; in the other two cases, the computation has no output value. The partial function which assigns this output value to the given input whenever it exists and which is undefined otherwise is called the function computed by the program.

We say that a partial function is RASP-computable iff it is the function computed by some RASP program.

Examples

(1) The following program computes the function $f(x,y) = x + y$. The program is stored in R1, R2, and R3; the input registers are R4 and R5.

9. RANDOM-ACCESS STORED PROGRAM MACHINES 177

R1	CLA⟨$\bar{4}$⟩
R2	ADD⟨$\bar{5}$⟩
R3	HALT
R4	(input x)
R5	(input y)

The story of the computation is as follows: First, the content of R4, say x, is stored in AC; to this, then, is added the content of R5, say y. Hence AC now contains $x + y$. The next instruction directs the machine to halt; at that moment AC contains $x + y$, as it should.

(2) The following program computes the function $f(x,y) = x \cdot y$. The program is stored in R1 to R10; the input registers are R11 and R12.

R1	CLA⟨$\bar{12}$⟩
R2	TRZ $\bar{9}$
R3	SUB $\bar{1}$
R4	STO $\bar{12}$
R5	CLA $\bar{13}$
R6	ADD⟨$\bar{11}$⟩
R7	STO $\bar{13}$
R8	TRA $\bar{1}$
R9	CLA⟨$\bar{13}$⟩
R10	HALT
R11	(input x)
R12	(input y)
R13	$\bar{0}$

This program keeps an intermediate sum register R13. At the outset, this register has content 0, and the input registers R11 and R12 have contents x and y, respectively. The program computes $x \cdot y$ by adding x to itself y times.

(3) Next consider the following program in registers R1 to R16 with input register R17.

R1	CLA⟨$\bar{18}$⟩
R2	TRZ $\bar{11}$
R3	SUB $\bar{1}$
R4	STO $\bar{18}$
R5	CLA⟨$\bar{15}$⟩
R6	STO⟨$\bar{17}$⟩

R7	CLA⟨$\overline{17}$⟩
R8	ADD $\overline{1}$
R9	STO $\overline{17}$
R10	TRA $\overline{1}$
R11	CLA⟨$\overline{16}$⟩
R12	STO⟨$\overline{17}$⟩
R13	CLA $\overline{0}$
R14	TRA $\overline{19}$
R15	ADD $\overline{5}$
R16	HALT
R17	$\overline{19}$
R18	(input x)

This may be a bit hard to see, but the program above computes nothing else but $5 \cdot x$ for any input $x \geqslant 1$. It does go about it in a somewhat roundabout fashion. Namely, after reading the content x of R18, it then proceeds to write "ADD 5" into each one of the registers R19 to R$(19+x-1)$ and "HALT" into R$(19+x)$. After that, it transfers control to R19 and thus proceeds to add 5 x times to the content of AC and then halts.

Observe that the RASP program in example (3) differs from the others in the following respect: In examples (1) and (2), control is never transferred to a register whose content was ever changed during the computation. Such programs are called *fixed programs*. In contrast, example (3) shows a program whose execution involves the execution of instructions that are themselves found and stored in appropriate registers during the computation itself. Such programs are called *self-modifying*. Of course, it is hard to tell at first glance whether a given RASP program is self-modifying or not. This decision involves a knowledge of the courses a computation can take. Indeed, the reader will show in a later problem that the problem is recursively undecidable.

9.1 *Problems*

(a) Write a fixed RASP program for the computation of $f(x) = [x/5]$, the integral part of $x/5$.

(b) Write a self-modifying RASP program for the computation of $g(x,y) = [x/y]$.

(c) Discuss the function computed by the following RASP program:

9. RANDOM-ACCESS STORED PROGRAM MACHINES   179

R1	CLA⟨$\bar{8}$⟩
R2	SUB⟨$\bar{9}$⟩
R3	TRZ $\bar{6}$
R4	STO $\bar{8}$
R5	TRA $\bar{1}$
R6	CLA⟨$\bar{8}$⟩
R7	HALT
R8	(input x)
R9	(input y)

(d) Discuss the function computed by the following RASP program:

R1	CLA⟨$\bar{9}$⟩
R2	STO $\overline{12}$
R3	STO $\overline{13}$
R4	STO $\overline{14}$
R5	CLA⟨$\overline{10}$⟩
R6	STO $\overline{15}$
R7	CLA $\bar{0}$
R8	TRA $\overline{12}$
R9	ADD⟨$\overline{11}$⟩
R10	HALT
R11	(input x)

FUNCTIONS COMPUTABLE ON RASP

We approach the problem as to what functions are computable on RASP by comparing RASP computations with computations on the universal calculator. Let there be given a program π for the universal calculator. Can we find a RASP program that computes the same function?

Clearly, every individual operation of π can be mimicked on RASP. For example,

$$\underline{do}\ x_k := S(x_k)\ \underline{then}\ldots$$

would be performed by first collecting the value of x_k, stored perhaps in R_{k*}, into the accumulator, then add 1, then store the result back into R_{k*}. Thus, the relevant piece of RASP program would be

CLA$\langle \overline{k*} \rangle$
ADD $\overline{1}$
STO $\overline{k*}$.

The other operational instructions are performed similarly. Even the decision step in the conditional instruction

$$\underline{\text{if }} x_k = 0 \underline{\text{ then}} \ldots$$

can easily be performed by the available conditional transfer

$$\text{TRZ} \ldots .$$

The nontrivial step of passage from π to a RASP program concerns the go to directions in π. This is the main step in the proof of the following result.

9.2 Theorem. For every program π of the universal calculator, we can effectively find a (fixed) RASP program π^* such that π and π^* compute the same function.

Proof. Without loss of generality, we may assume that the function computed by π is understood as the value of x_1 at termination. Let us now describe how π^* is constructed from π.

If π has only start and halt instructions, then it must be of the form

$$\underline{\text{start}}: \underline{\text{go to}} \ 1,$$
$$1: \underline{\text{halt}}.$$

The corresponding RASP program π^* is then simply

R1   CLA$\langle \overline{3} \rangle$
R2   HALT
R3   (input x_1)

and R3 is understood to be the input register corresponding to input x_1.

Suppose that π is any program for the universal calculator with m instructions (not counting the start instruction), m_0 of which are not halt instructions, and suppose that the variables contained in π are

9. RANDOM-ACCESS STORED PROGRAM MACHINES

among $x_1, \ldots, x_n$. Suppose that these instructions are labeled $1, 2, \ldots, m$. Then there are numbers

$$1^*, \ldots, m^* \quad \text{and} \quad 1^{**}, \ldots, n^{**}$$

and a RASP program π^* such that π^* results from π by replacing each line

$$k: \ldots$$

of π by one, two, three, or four consecutive lines according to Table 2.2. The numbers $1^*, \ldots, m^*$ indicate the line on which that piece of π^* starts that corresponds to the instruction with label $1, \ldots, m$. The numbers $1^{**}, \ldots, n^{**}$ are the first n numbers such that $R1^{**}, R2^{**}, \ldots$ do not contain instructions of the program π^*. These can then serve as input registers.

TABLE 2.2

Line of π	Replacement lines in π^*
start: go to p	TRA $\overline{p^*}$
k: do $x_s := S(x_s)$ then go to p	CLA $\langle \overline{s^{**}} \rangle$
	ADD $\overline{1}$
	STO $\overline{s^{**}}$
	TRA $\overline{p^*}$
k: do $x_s := P(x_s)$ then go to p	CLA $\langle \overline{s^{**}} \rangle$
	SUB $\overline{1}$
	STO $\overline{s^{**}}$
	TRA $\overline{p^*}$
k: do $x_s := x_r$ then go to p	CLA $\langle \overline{r^{**}} \rangle$
	STO $\overline{s^{**}}$
	TRA $\overline{p^*}$
k: do $x_s := 0$ then go to p	CLA $\overline{0}$
	STO $\overline{s^{**}}$
	TRA $\overline{p^*}$
k: go to p	TRA $\overline{p^*}$
k: if $x_r = 0$ then go to p else go to q	CLA $\langle \overline{r^{**}} \rangle$
	TRZ $\overline{p^*}$
	TRA $\overline{q^*}$
k: halt	CLA $\langle \overline{1^{**}} \rangle$
	HALT

2. RECURSIVE FUNCTIONS AND PROGRAMMED MACHINES

It requires but a little thought on the part of the reader to convince himself that π^* indeed computes the same function as π does. Note that π^* is a fixed program.

Example. Consider the program π for the universal calculator

start: go to 1;
  1: do $x_1 := x_3$ then go to 2;
  2: if $x_2 = 0$ then go to 5 else go to 3;
  3: do $x_2 := P(x_2)$ then go to 4;
  4: do $x_1 := S(x_1)$ then go to 2;
  5: halt.

The corresponding program π^* for RASP is found as follows:

$$
\begin{array}{lllll}
(1) & \left\{\begin{array}{l} \text{CLA}\langle\overline{3^{**}}\rangle = \text{CLA}\langle\overline{19}\rangle \\ \text{STO } \overline{1^{**}} = \text{STO } \overline{17} \\ \text{TRA } \overline{2^*} = \text{TRA } \overline{4} \end{array}\right. & : & \begin{array}{l} \text{R1} \\ \text{R2} \\ \text{R3} \end{array} \\[2em]
(2) & \left\{\begin{array}{l} \text{CLA}\langle\overline{2^{**}}\rangle = \text{CLA}\langle\overline{18}\rangle \\ \text{TRZ } \overline{5^*} = \text{TRZ } \overline{15} \\ \text{TRA } \overline{3^*} = \text{TRA } \overline{7} \end{array}\right. & : & \begin{array}{l} \text{R4} \\ \text{R5} \\ \text{R6} \end{array} \\[2em]
(3) & \left\{\begin{array}{l} \text{CLA}\langle\overline{2^{**}}\rangle = \text{CLA}\langle\overline{18}\rangle \\ \text{SUB } \overline{1} = \text{SUB } \overline{1} \\ \text{STO } \overline{2^{**}} = \text{STO } \overline{18} \\ \text{TRA } \overline{4^*} = \text{TRA } \overline{11} \end{array}\right. & : & \begin{array}{l} \text{R7} \\ \text{R8} \\ \text{R9} \\ \text{R10} \end{array} \\[2em]
(4) & \left\{\begin{array}{l} \text{CLA}\langle\overline{1^{**}}\rangle = \text{CLA}\langle\overline{17}\rangle \\ \text{ADD } \overline{1} = \text{ADD } \overline{1} \\ \text{STO } \overline{1^{**}} = \text{STO } \overline{17} \\ \text{TRA } \overline{2^*} = \text{TRA } 4 \end{array}\right. & : & \begin{array}{l} \text{R11} \\ \text{R12} \\ \text{R13} \\ \text{R14} \end{array} \\[2em]
(5) & \left\{\begin{array}{l} \text{CLA}\langle\overline{1^{**}}\rangle = \text{CLA}\langle\overline{17}\rangle \\ \text{HALT} = \text{HALT} \end{array}\right. & : & \begin{array}{l} \text{R15} \\ \text{R16} \end{array} \\[1em]
& \begin{array}{l} (\text{input } x_1) \\ (\text{input } x_2) \\ (\text{input } x_3) \end{array} & : & \begin{array}{l} \text{R17} \\ \text{R18} \\ \text{R19} \end{array}
\end{array}
$$

The middle column is the desired program.

9. RANDOM-ACCESS STORED PROGRAM MACHINES

Remark. The passage from π to π^* is called "compiling" in the jargon of computer programmers. A program which performs this task automatically is called a "compiler." We are not writing such a program here.

Let us now ask the converse question, namely: Can we find for every RASP program a program for the universal calculator which computes the same function? It would be rather surprising if we were unable to do that, because we have convinced ourselves intuitively in this chapter that all "computable functions," and hence certainly all RASP-computable functions, are partial recursive (and hence computable on the universal calculator). Let us prove this fact directly.

9.3 Theorem. Every RASP-computable function is computable on the universal calculator.

Proof. The simplest way to prove this is to realize the program of p on the universal calculator. To do this in full detail would be neither feasible here nor illuminating. Let us therefore be content with a sketch of the necessary steps that are involved in such a realization.

First, we need to select among the set of memory cells of the universal calculator one cell corresponding to AC, one corresponding to IC, and one containing an encoding of the sequence of contents of R1, R2, ... In addition, we shall want to reserve some more cells for "working space."

Next we encode the possible contents of RASP registers, and sequences of such contents, as numerals in some efficient way (e.g., using prime factorization). The central step, then, consists of writing subroutines for the various procedures such as

$$\text{test } R\langle IC \rangle \begin{array}{c} \text{TRA } \bar{n} \\ \vdots \\ \text{HALT} \end{array},$$

$$\longrightarrow \langle AC \rangle := \langle Rn \rangle \longrightarrow,$$

and

$$\longrightarrow \langle AC \rangle := \langle AC \rangle + \langle Rn \rangle \longrightarrow.$$

The possibility of doing this has already been discussed for a very similar task in Section 8.

Now, suppose we are given a RASP program π^* in registers R1, ..., Rm and with input registers R$m+1$, ..., R$m+n$. Then we consider the following program π for the universal calculator: The program π first collects the contents of its memory cells $x_1, ..., x_n$ and produces the code for the program π^* followed by the values of $x_1, ..., x_n$. This number is then put into the memory cell that is designated above for the purpose of containing the encoding of the sequence of contents of R1, R2, ... After this it starts the execution of the program of Fig. 2.66 (with the individual boxes replaced by the appropriate subroutines as described above).

THE TIME ADVANTAGE OF SELF-MODIFYING PROGRAMS[10]

Let us restrict attention to functions of one variable only. For such functions, we may redefine the notion of RASP computability by assuming that the input is already in the accumulator at the outset of the computation.

9.4 Lemma. If f is computable on RASP by a program π in $G(n)$ steps (not counting TRA and TRZ) then, for all sufficiently large n,

$$f(n) \leq n \cdot 3^{[G(n)/3]}$$

(where $[K/3]$ is the integral part of $K/3$).

Proof. Let s be the maximal number stored in any register R_j at the start of the program for $f(n)$. Let m be the number of instructions of the form ADD ... that are executed before we arrive at the first instruction of the form STO. Observe that at this moment the content of the accumulator is at most $r + m \cdot s$. Could we have done better in this number of steps? Let us assume $m \geq 3 \cdot k$ and consider a sequence of instructions

[10] This material is due to Hartmanis [11].

$$\underbrace{\text{STO}\,\bar{p};\,\text{ADD}\langle\bar{p}\rangle;\,\text{ADD}\langle\bar{p}\rangle}_{(1)};\,\underbrace{\text{STO}\,\bar{p};\,\text{ADD}\langle\bar{p}\rangle;\,\text{ADD}\langle\bar{p}\rangle}_{(2)};\,\ldots;\,\underbrace{\ldots;\,\text{ADD}\langle\bar{p}\rangle}_{(k)}.$$

It is obvious that this sequence produces an output of $n \cdot 3^k$. Now, if $n > 2 \cdot s$, then $n \cdot 3^k > n + 3k \cdot s$ for all k as is easily verified. The moral is, that if we want a program for which the value of the accumulator grows as fast as possible with the number of steps, we best look for a program of the form

$$\underbrace{\text{STO};\,\text{ADD};\,\ldots;\,\text{ADD}}_{m_1 - 1};\,\underbrace{\text{STO};\,\text{ADD};\,\ldots;\,\text{ADD}}_{m_2 - 1},\,\underbrace{\text{STO};\,\text{ADD};\,\ldots;\,\text{ADD}}_{m_3 - 1};\,\ldots.$$

Such a program increases the initial value n of the accumulator in $m_1 + m_2 + \cdots$ steps to $n \cdot (m_1 \cdot m_2 \cdots)$. Next we remark that certain sequences

$$\text{STO};\,\text{ADD};\,\ldots;\,\text{ADD}$$

make the accumulator grow faster than others. Namely:

(a) If we compare k_3 pieces of the form STO; ADD; ADD with k_2 pieces of the form STO; ADD and if $k_3 \cdot 3 = k_2 \cdot 2$, then the sequence consisting of STO; ADD; ADD produces the bigger value. The sequence of k_3 triples STO; ADD; ADD will produce $n \cdot 3^{k_3}$ while the sequence of k_2 pairs STO; ADD will produce $n \cdot 2^{k_2}$. But observe that if $k_3 \cdot 3 = k_2 \cdot 2$, then $n \cdot 2^{k_2} < n \cdot 3^{k_3}$. Namely $2^{k_2} = 2^{k_3 \cdot 3/2} < 3^{k_3}$ because $2^{3/2} < 3$.

(b) By the same kind of argument we can show that if we take $m > 3$ and k_m pieces of the form

$$\text{STO};\,\underbrace{\text{ADD};\,\ldots;\,\text{ADD}}_{m - 1}$$

and if $k_m \cdot m = k_3 \cdot 3$, the sequence of triples again produces faster growth. So altogether, if n is sufficiently large, the fastest growing function that can be obtained is by a sequence of triples STO; ADD; ADD, which produces $n \cdot 3^k$ in $3k$ steps.

9.5 Corollary. Let g be any function. Then every RASP program for $f(n) = n \cdot 3^{g(n)}$ takes at least $3 \cdot g(n)$ steps (not counting TRA and TRZ) to compute $f(n)$.

Proof. By Lemma 9.4, we have $f(n) = n \cdot 3^{g(n)} \leq n \cdot 3^{[G(n)/3]}$, and hence $g(n) \leq [G(n)/3]$ and $3 \cdot g(n) \leq G(n)$.

To illustrate the phenomenon that self-modifying programs are faster than all fixed programs (for some functions), let us consider the following function

$$f(n) = n \cdot 3^{(n^2)}.$$

This function can be computed by a self-modifying program π, described below, which is the fastest program for f in a certain sense, also specified below. Moreover, we can even predict by how much slower *any* fixed program for f must be.

By *computing time* $F(n)$ of a function $f(n)$ by a program π, we understand the number of steps (including TRA and TRZ) that the computation according to π takes for input n.

9.6 Theorem. There is a RASP program π for $f(n) = n \cdot 3^{n^2}$ with computing time $F(n)$ such that no RASP program computes f in time $(1 - \varepsilon) \cdot F(n)$ for any $\varepsilon > 0$. Moreover, every *fixed* RASP program π' of length l has a computing time $F'(n)$ such that $F'(n) \geq \bigl(1 + (1/2l)\bigr) \cdot F(n)$ for all sufficiently large n.

Proof. Let us first describe the program π. For given n, the first task of the program is to produce a sequence of instructions of the form

$$
\begin{array}{l}
Rr: \text{CLA}\langle \bar{p} \rangle \\
\quad \text{STO } \bar{k} \\
\quad \text{ADD}\langle \bar{k} \rangle \\
\quad \text{ADD}\langle \bar{k} \rangle \\
\quad \text{STO } \bar{k} \\
\quad \text{ADD}\langle \bar{k} \rangle \\
\quad \text{ADD}\langle \bar{k} \rangle \\
\end{array}
\begin{array}{l}
\\
\left.\begin{array}{l}\\ \\ \\\end{array}\right\} (1) \\
\left.\begin{array}{l}\\ \\ \\\end{array}\right\} (2) \\
\end{array}
$$

$$\vdots$$
$$\left.\begin{array}{l}\text{STO } \bar{k} \\ \text{ADD}\langle \bar{k}\rangle \\ \text{ADD}\langle \bar{k}\rangle\end{array}\right\} (n)$$
STO $\bar{q}$
CLA$\langle \bar{p}\rangle$
SUB $\bar{1}$
TRZ $\bar{s}$
STO $\bar{p}$
CLA$\langle \bar{q}\rangle$
TRA $\bar{r}$
Rs: CLA$\langle \bar{q}\rangle$
HALT
Rk: $\bar{0}$
Rp: $\bar{n}$
Rq: $\bar{0}$

After it has filled registers Rr through Rq with these data, the program transfers control to register Rr. The ensuing computation is nothing but the execution of the piece of program listed above. When it finally halts, the value in the accumulator will be $n \cdot 3^{n^2}$. Namely, each triple of instructions

$$\text{STO } \bar{k}$$
$$\text{ADD } \langle \bar{k}\rangle$$
$$\text{ADD } \langle \bar{k}\rangle$$

triples the content of AC. Hence, after running through the sequence of STO; ADD; ADD triples once, the content of the accumulator will be $n \cdot 3^n$. The remaining instructions direct this sequence to be run through n times. Therefore, when HALT is finally reached, the content of AC will be $n \cdot (3^n)^n = n \cdot 3^{n^2}$.

We leave the actual writing of the program π to the imagination of the reader. However, from the description of π above, we can still discuss the computing time of π. Namely, it is clear that we can find a constant c such that for any input n it takes at most $c \cdot n$ steps to write the piece of program indicated above. The execution of that piece of program itself takes at most $(3 \cdot n + d) \cdot n$ steps for some other, easily determined, constant d. Altogether, we therefore see that

$$F(n) \leq 3n^2 + d \cdot n + c \cdot n = 3n^2 + e \cdot n$$

for some fixed constant e.

Assume now that there were a RASP program π' with computing time $F'(n) \leq (1 - \varepsilon) F(n)$ for some $\varepsilon > 0$. By the corollary above, we know that $F'(n) \geq 3n^2$, since even if we count only steps different from TRA and TRZ, the program π' must take at least $3n^2$ steps. However, let us consider numbers $n > e/3\varepsilon$ and use our estimate of $F(n)$ above. It follows that

$$F'(n) \leq (1 - \varepsilon) F(n) \leq \left(1 - \frac{e}{3n}\right) \cdot (3n^2 + e \cdot n) = 3n^2 - \frac{e^2}{3} < 3n^2,$$

which is a contradiction, proving the first part of the theorem.

To prove the second part of the theorem, let us consider a RASP program π' for $f(n) = n \cdot 3^{n^2}$. Assume that π' is a *fixed* program consisting of l instructions. By the corollary, the execution of π' for input n will involve at least $3n^2$ operations which are not of the form TRA or TRZ. Also, control cannot ever leave the l registers in which the program is stored. Hence, every lth step at least will be a transfer TRA or TRZ of control. Hence π' will execute at least $3n^2/l$ such transfer steps. Thus, altogether

$$F'(n) \geq 3n^2 + \frac{3n^2}{l} = 3n^2 \frac{(l + 1)}{l}.$$

By our result above, we have

$$\left(1 + \frac{1}{2l}\right) \cdot F(n) \leq \left(1 + \frac{1}{2l}\right)(3n^2 + e \cdot n) = 3n^2 + e \cdot n + \frac{n^2}{2l} + \frac{e \cdot n}{2l}$$

$$= \left(1 + \frac{1}{l}\right) \cdot 3n^2 - \frac{3n^2}{2l} + en + \frac{e \cdot n}{2 \cdot l}.$$

This last expression is strictly smaller than

$$\left(\frac{l+1}{l}\right) \cdot 3n^2$$

if $n^2/2l > e \cdot n + en/2l$, which is the case iff $n > (2 \cdot e \cdot l/3) + (e/3)$. Thus, for n sufficiently large, we have the following inequality:

$$\left(\frac{l+1}{2l}\right) \cdot F(n) < \left(\frac{l+1}{l}\right) \cdot 3n^2.$$

If we combine this with the inequality

$$3n^2 \left(\frac{l+1}{l}\right) \leq F'(n)$$

proved above, we get the result that

$$\left(1 + \frac{1}{2 \cdot l}\right) \cdot F(n) < F'(n)$$

for all sufficiently large n.

REFERENCES

Books and Collections

[1] M. Davis, *Computability and Unsolvability*. McGraw-Hill, New York, 1958.
[2] M. Davis (ed.), *The Undecidable*. Raven, New York, 1965.
[3] H. Hermes, *Enumerability, Decidability, Computability*. Springer Publ., New York, 1965.
[4] S. C. Kleene, *Introduction to Metamathematics*. Van Nostrand-Reinhold, Princeton, New Jersey, 1952.
[5] A. A. Markov, *The Theory of Algorithms* (Engl. transl.). U.S. Dept of Commerce, 1962.
[6] M. Minsky, *Computation: Finite and Infinite Machines*. Prentice-Hall, Englewood Cliffs, New Jersey, 1967.

[7] R. Peter, *Recursive Functions*. Academic Press, New York, 1967.
[8] H. Rogers, *Theory of Recursive Functions and Effective Computability*. McGraw-Hill, New York, 1967.
[9] A. Yasuhara, *Recursive Function Theory and Logic*. Academic Press, New York, 1971.

Original Articles

[10] C. C. Elgot and A. Robinson, Random-access stored-program machines, an approach to programming languages. *J. Assoc. Comput. Mach.*, **11**, 365–399 (1964).
[11] J. Hartmanis, Computational complexity of random access stored program machines. *Math. Systems Theory*, **5**, 232–245 (1971).
[12] S. C. Kleene, General recursive functions of natural numbers. *Math. Ann.*, **112**, 727–742 (1936).
[13] A. R. Meyer and D. M. Ritchie, Computational complexity and program structure. *IBM Res. Rep.*, 1817 (1967).
[14] J. C. Shepherdson and H. E. Sturgis, The computability of partial recursive functions. *J. Assoc. Comput. Mach.*, **10**, 217–255 (1963).
[15] A. Turing, On computable numbers, with an application to the Entscheidungsproblem. *Proc. London Math. Soc.* (2), **42**, 230–265 (1937).
[16] H. Wang, A variant to Turing's theory of computing machines. *J. Assoc. Comput. Mach.*, **4**, 63–92 (1957).

CHAPTER

Elementary Syntax

In formal languages such as the ones encountered in mathematical logic and in computer programming, we make a distinction between syntactic and semantic aspects.

The *syntax* of a language is concerned with the actual symbolic presentation of expressions of the language. Various devices, such as punctuation (commas, semicolons, colons, parentheses, brackets, braces, and quotation marks), typographical conventions (italics, boldface), and special symbols, are employed to enable the reader to decipher and understand the expressions. Already in nonformal texts, these syntactical devices play a considerable role in making the meaning of a sentence clear. The degree of sophistication in the notation needs, however, to be rather low if the reader is an educated human being. (Indeed, too much of it is considered bad style.) Then we can appeal to context and good will of the reader if perhaps the text is slightly opaque or ambiguous. No such recourse is possible, however, if the reader of the text is a machine. Machines are singularly literal minded and devoid of imagination. Thus considerable attention has to be given to syntactical questions in languages—such as programming languages—that are designed for the benefit of the computer.

The *semantics* of a formal language concerns itself with the meaning, that is, the intended interpretation, of expressions of the language.

3. ELEMENTARY SYNTAX

Consider an algebraic expression such as

$$a \cdot (b + c) de + (f + g)(h + k).$$

Every student is able to correctly interpret this algebraic expression. He finds this task so simple even, that he sees little need to pause and think about what is actually involved in the process of reading. This is as it should be. But if the expression above it is part of an instruction to a computer, as it may well be, then the matter of interpretation becomes one of mathematical concern.

The structure of the algebraic expression set above is best organized as a tree, called its *parsing* tree:

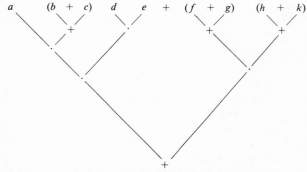

And it appears at least slightly more obvious how one would translate such a tree into a program (for the evaluation of the expression) than how one would translate the original expression. Clearly, there is more than one parsing tree for this algebraic expression. For example,

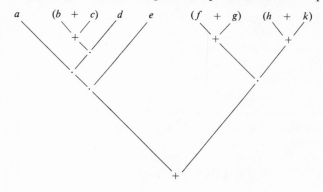

1. THE STRUCTURE OF LANGUAGE

The reason is that, while multiplication is a binary operation, we have used its associativity to say, in effect, that we do not care whether abc is understood as $(ab)c$ or as $a(bc)$. Thus we have introduced an ambiguity into the notation, one which does not matter here, but could obviously be destructive of meaning in other circumstances.

For example, consider a programming language which allows instructions to form by two types of conditionalizations:

$$\underline{\text{if}}\ \alpha\ \underline{\text{then}}\ \beta$$

$$\underline{\text{if}}\ \alpha\ \underline{\text{then}}\ \beta\ \underline{\text{else}}\ \gamma$$

where α can again be replaced by such a conditional instruction, and so on. How should we read the instruction

$$\underline{\text{if}}\ \alpha_1\ \underline{\text{then}}\ \underline{\text{if}}\ \alpha_2\ \underline{\text{then}}\ \beta_1\ \underline{\text{else}}\ \beta_2\ ?$$

There are two parsing trees, namely

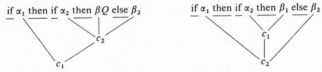

where c_1 is the conditionalization of the first type (two arguments), and c_2 is the conditionalization of the second type (three arguments). This time, the ambiguity is rather serious and a computer language in which such a feature is built in would be quite unreasonable.

Our goal in the present chapter is to investigate the mathematical background of notation systems and to develop methods which will allow us to formulate, and deal with, problems of generation and translation of such systems, in particular in the context of algorithmic languages.

1. THE STRUCTURE OF LANGUAGE

The two central concepts with which we are concerned here are those of *syntactical operation* and *syntactical category*. Let us consider the algorithmic language given by the capabilities

$$x := 0, \quad x := 1, \quad x := S(x), \quad x := P(x), \quad x = 0$$

3. ELEMENTARY SYNTAX

and let us consider programs over these capabilities as strings of symbols over a finite alphabet. Thus the following alphabet is sufficient:

$\{x, S, P, :, \cdot :=, \underline{\text{start}}, \underline{\text{halt}}, \underline{\text{do}}, \underline{\text{then}}, \underline{\text{go to}}, \underline{\text{if}}, \underline{\text{else}}, 0, 1, ;, (,)\}$.

By underlining start, and so on, we indicate that we wish to consider these assemblages as single symbols. The following syntactical categories are distinguished:

Labels: smallest set of words containing 1 and closed under the syntactical operations f_0 and $f_1: f_0(w) = w0, f_1(w) = w1$.

Operations: $\{x := 0, x := 1, x := S(x), x := P(x)\}$.

Operational instructions: set of all words of the form $k: \underline{\text{do}}\ \alpha\ \underline{\text{then}}\ \underline{\text{go to}}\ p$, where α belongs to the syntactical category of operations and k and p belong to the syntactical category of labels. (What we have here is a ternary syntactical operation **op** whose first argument ranges over labels, whose second argument ranges over operations, and whose third argument ranges over labels again. The result of this syntactical operation lies in the syntactical category of operational instructions.)

Conditional instructions: set of words of the form $k: \underline{\text{if}}\ \beta\ \underline{\text{then}}\ \underline{\text{go to}}\ p$ $\underline{\text{else}}\ \underline{\text{go to}}\ q$ where β is of the form $x = 0$ and p, q, and k are labels. (Again we have a ternary syntactical operation **con**.)

Halt instructions: set of words of the form $k: \underline{\text{halt}}$, k being a label. (This time we have a unary syntactical operation, **term**.)

Program bodies: smallest set of words that contains the set of operational, conditional, and halt instructions and closed under the operations **seq**, where $\mathbf{seq}(w_1, w_2)$ is defined if w_2 is a program body, w_1 an instruction. Then $\mathbf{seq}(w_1, w_2)$ is the word $w_1; w_2$.

Programs: set of words of the form $\underline{\text{start}}: \underline{\text{go to}}\ k; w$, where w is a program body. (We have a binary syntactical operation present here which we denote by **init**.)

Remark. The alert reader may notice that not all programs herewith defined would have been accepted as such in the preceding chapter. More about this later.

1. THE STRUCTURE OF LANGUAGE

The basic concern is that we would like to be able to *read* a program and understand its structure. Thus, as a first step, given a program, we should like to see whether a word π is a program, and if it is, what are the two words k and w such that $\pi = \mathbf{init}(k,w)$. After that we would go on with w in the same manner, and so forth.

Let us now, for a moment, abstract from the particular language present and seek to perceive the general pattern. Consider the following definition.

1.1 Definition. A *grammatical structure* is a structure of the form $G = \langle C_1, \ldots, C_m; g_1, \ldots, g_k \rangle$ where the C_i are nonempty, mutually exclusive sets, called syntactical categories, and each g_i is a map from $C_{k_1} \times \cdots \times C_{k_r}$ to C_l for some $k_1, \ldots, k_r, l \leq m$, subject to the following two axioms:

(i) *Unique readability.* If $g_i(u_1, \ldots, u_r) = g_j(v_1, \ldots, v_s)$, then $i = j$, $r = s$, and $u_1 = v_1, \ldots, u_r = v_s$.

(ii) *Induction.* If $D \subseteq C_1 \cup \cdots \cup C_m$ contains all elements that are not of the form $g_i(w_1, \ldots, w_r)$ for any $i, r,$ and $w_1, \ldots, w_r$ (called *atoms* of G) and is closed under all operations g_j, then $D = C_1 \cup \cdots \cup C_m$.

The first axiom simply expresses the property of nonambiguous interpretability of properly formed words. The second axiom plays a role similar to the induction axiom for free semigroups; it ensures the fact that the structure G has all the mathematical features of a system of words over some alphabet. This is also the axiom which usually is the most easily verified; in the case of the programming language above, it reduces to the fact that we have indeed taken the smallest sets of words containing certain given words and closed under the various syntactical operations. Thus, the programming language introduced above can be viewed as a structure

$$\langle \text{labels, operations}, \ldots, \text{programs}; f_0, f_1, \mathbf{op}, \ldots, \mathbf{seq}, \mathbf{init} \rangle.$$

Both axioms are easily verified.

1.2 Definition. Let $G = \langle C_1, \ldots, C_m; g_1, \ldots, g_n \rangle$ be an algebraic structure which satisfies axiom (ii) [but not necessarily axiom (i)] of grammatical structures, and let $w \in C_1 \cup \cdots \cup C_m$. A *parsing tree* for w is a finite tree whose nodes are labeled by symbols $g_1, \ldots, g_n$ and whose leaves are labeled by atomic elements of G in such a manner that there exists an assignment of elements of $C_1, \ldots, C_m$ to the nodes which satisfies the following conditions:

(i) The root of the tree is assigned w.
(ii) If a node with a label g_i is assigned u and its immediate predecessors are assigned $v_1, \ldots, v_s$ (in that order), then g_i is s-ary and $u = g_i(v_1, \ldots, v_s)$.

Example. Let $N = \langle N, +, \cdot \rangle$ be the set of natural numbers greater than or equal to 2, under addition and multiplication. This structure clearly satisfies axiom (ii); the only atomic elements are 2 and 3. Then the following is a parsing tree of the number 128:

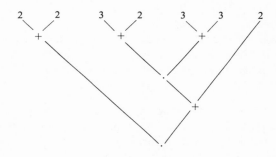

namely,

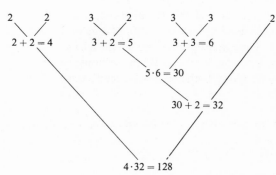

Of course, there are other parsing trees for 128. (How many?)

2. INTRODUCTION TO THE THEORY OF CONTEXT-FREE GRAMMARS

1.3 Proposition. A structure $\langle C_1,...,C_m; g_1,...,g_n \rangle$ which satisfies axiom (ii) is *grammatical* iff each element of $C_1 \cup \cdots \cup C_m$ has a unique parsing tree.

The proof is obvious, using axiom (ii).

1.4 Problems
(a)* Show that the following structure is grammatical:

$$L = \langle L; \wedge, \vee, \neg \rangle,$$

where L is the smallest set of words over K, A, N, p, q, r which contains p, q, r and is closed under the operations

$$\wedge(w_1, w_2) = Kw_1 w_2, \qquad \vee(w_1, w_2) = Aw_1, w_2, \qquad \neg(w) = Nw.$$

(This is the so-called Polish notation system for propositional logic invented by Łukasiewicz; it can be used, with obvious changes, as a notation system for any grammatical structure whatever.)

(b)* Generalize Theorems 2.4 and 2.5 of Chapter 1 from free monoids to grammatical structures.

2. INTRODUCTION TO THE THEORY OF CONTEXT-FREE GRAMMARS

We now turn to a more familiar way of looking at a language (such as the computer programs of Section 1), namely as languages generated by grammar in the fashion of Chapter 1, Section 4. Only this time we have to generalize the notion of a grammar, from the right-linear grammars considered there, to what we shall call context-free grammars. Otherwise, as we see presently, we would not obtain all the languages that we wish to consider. We start with the general definition of a grammar. After this we make the connection between grammars and the grammatical structures of the previous section. This connection in turn then motivates the basic questions which are treated in the theory of context-free grammars.

2.1 Definition

(a) A *grammar* G over a terminal alphabet T consists of a finite set N of nonterminal symbols, disjoint from T, and a finite set P of productions. Each production is in the form

$$u \to v$$

where u and v are any words over the combined alphabet $T \cup N$; u is nonempty. (Note that in the case of right-linear grammars we have asked these productions to take on particularly simple forms!)

(b) If x and y are words over $T \cup N$, we write

$$x \underset{P}{\Rightarrow} y$$

(or simply $x \Rightarrow y$ if P is understood), if there are words w_1 and w_2 in $(T \cup N)^*$ and a production $u \to v$ in P such that

$$x = w_1 u w_2 \quad \text{and} \quad y = w_1 v w_2.$$

(In other words, a piece of x which is of the form u is being replaced by v.)

(c) We write $x \underset{P}{\overset{*}{\Rightarrow}} y$ if $x = y$ or there is a finite sequence $x_0, x_1, x_2, \ldots, x_n$ such that $x = x_0$, $x_n = y$, and

$$x_{i-1} \underset{P}{\Rightarrow} x_i$$

for $i = 1, 2, \ldots, n$.

(d) Let S be any nonterminal symbol in N. The set of words w over the terminal alphabet T for which

$$S \underset{P}{\overset{*}{\Rightarrow}} w$$

is called the set generated (or produced) by the grammar G starting at S.

To orient ourselves, let us try to consider the programming language of Section 1 as the language generated by a grammar G. The terminal alphabet is obvious; as a nonterminal alphabet we choose

$$\{L, \Omega, C, H, B, \Pi\}.$$

2. INTRODUCTION TO THE THEORY OF CONTEXT-FREE GRAMMARS

(Note that we have introduced a nonterminal symbol for each syntactic category.) We consider the following productions:

$$L \to L0$$
$$L \to L1$$
$$L \to 1$$
$$\Omega \to x := 0$$
$$\Omega \to x := 1$$
$$\Omega \to x := S(x)$$
$$\Omega \to x := P(x)$$
$$0 \to L: \underline{\text{do}}\ \Omega\ \underline{\text{then go to}}\ L$$
$$C \to L: \underline{\text{if}}\ x = 0\ \underline{\text{then go to}}\ L\ \underline{\text{else go to}}\ L$$
$$H \to L: \underline{\text{halt}}$$
$$B \to 0$$
$$B \to C$$
$$B \to H$$
$$B \to 0; B$$
$$B \to C; B$$
$$B \to H; B$$
$$\Pi \to \underline{\text{start: go to}}\ L; B.$$

The set of words produced starting at L are the labels; starting at Ω we produce operations; and so on. Programs are produced when we start at Π.

Remarks. (1) We observe that at least one part of the production system, namely that concerned with producing labels only, is a *right linear* grammar. Thus the set of labels forms a regular set. But also the remaining set of productions is of a rather simple form. Note, namely, that all productions are of the form $u \to v$ where *u consists of one non-terminal symbol only*. This means that in a production sequence, or "derivation," the nonterminal symbol u may be replaced by v *irrespective of the context* in which it occurs.

(2) The format of definition that is currently in use to present computer languages such as ALGOL is used below without any further explanations:

$$\langle\text{labels}\rangle ::= 1 \mid \langle\text{label}\rangle 1 \mid \langle\text{label}\rangle 0.$$

(3) Consider the following word:

start: go to 10; 11: halt.

This is clearly a program according to our present definition; indeed, it is generated as follows:

$\pi \Rightarrow$ start: go to L; $B \Rightarrow$ start: go to L; L: halt
$\Rightarrow$ start: go to $1L$; L: halt $\Rightarrow \cdots \Rightarrow$ start: go to 10; 11: halt.

But we would not accept this program since the go to instruction go to 10 is left dangling. We have violated a formation rule for programs *that is not embodied in our grammar*. Such global restriction rules are rather typical for programming languages and prevent them from falling altogether under our scheme.

It is natural to distinguish between various types of grammars according to the types of production rules that are allowed. Chomsky proposed the following four types of grammars for closer treatment.

2.2 Definition.[1] Let G be a grammar with terminal alphabet T, nonterminal alphabet N, and productions P.

(0) If there are no restrictions on the type of production allowed, then G is said to be a type-0 grammar.

(1) If each production is of the form $u \to v$ where v is at least the length of u, then G is said to be of type 1 or *context-sensitive*.

(2) If each production is of the form $A \to w$ where A is some nonterminal symbol and w is a nonempty word over $N \cup T$, then G is called *context-free* or of type 2.

---

[1] This definition is due to Chomsky; see, for example, Chomsky [3].

(3) If each production is of one of the forms $A \to aB$, $A \to b$, $A \to \lambda$, where A, B are nonterminal, and a, b are terminal symbols, then G is said to be of type 3 or *right-linear*.

We are quite familiar with grammars of type 3 from Chapter 1; we do not develop the theory of grammars of types 0 and 1 here, but only concentrate on grammars of type 2. The reason for this is that the step from context-free grammars to the grammatical structures of Section 1 (and vice versa) is particularly simple, which is finally also the reason why context-free grammars are of particular interest for the discussion of the structure and manipulation of programming languages.

Let G be a context-free grammar over the terminal alphabet T using a nonterminal alphabet $N = \{A_1, \ldots, A_m\}$. For each $A_i \in N$, let C_i be the class of words w over T such that $A_i \overset{*}{\Rightarrow} w$.

We shall use the sets C_i as the syntactic categories of the grammatical structure $\mathbf{G} = \langle C_1, \ldots, C_m, g_1, \ldots, g_n \rangle$ which we shall associate to the context-free grammar G. The operations $g_1, \ldots, g_n$ are associated to the productions of G as follows.

Let us consider a typical production in G:

$$A_i \to u_1 A_{k_1} u_2 A_{k_2} u \cdots u_r A_{k_r} u_{r+1}$$

where $u_j \in T^*$, $A_{k_u} \in N$, and $r \geqslant 0$. We associate to this production an r-ary syntactical operation

$$g: C_{k_1} \times C_{k_2} \times \cdots \times C_{k_r} \to C_i$$

defined by

$$g(v_1, \ldots, v_r) = u_1 v_1 u_2 v_2 u_3 \cdots u_r v_r u_{r+1}$$

(if $r = 0$, this is a constant function). Let $g_1, \ldots, g_n$ be an enumeration of all these syntactical operations associated to the grammar G.

2.3 Definition. The *structure associated* to a context-free grammar G is

$$\mathbf{G} = \langle C_1, \ldots, C_m; g_1, \ldots, g_m \rangle.$$

Is this structure grammatical? Referring back to Definition 1.1 we have to show that the C_i are nonempty, mutually exclusive sets and that the structure G satisfies the axioms of unique readability and induction. All we are given is the grammar G. Can we decide these questions "from simply looking at G"?

With respect to the emptiness problem we have the following result.

2.4 Proposition. It is decidable whether or not $C_i = \varnothing$.

Proof. Let

$$A_i = u_0 \Rightarrow u_1 \Rightarrow \cdots \Rightarrow u_s \Rightarrow u_{s+1} \Rightarrow \cdots \Rightarrow u_n = w$$

be a production sequence in G. We shall assign a depth indicator to each occurrence of a symbol in $u_0, u_1, \ldots, u_n$ according to the following rule: The one occurrence of A_i in u_0 is assigned the depth indicator 0. Suppose that u_{s+1} arises from u_s by a production $A_j \to v$; u_s is of the form $u'_s A_j u''_s$ and the depth indicators of occurrences of symbols in u_{s+1} are found as follows: All occurrence in u'_s and u''_s retain their old depth indicators; any occurrence of a symbol in v obtains the depth indicator $k + 1$ where k is the depth indicator of A_j in u_s. The depth of the derivation is the maximum of all depth indicators in w. Let m be the number of nonterminal symbols in P. If C_i is nonempty, then there is a derivation of a nonempty word from A_i which has depth at most m. Namely, suppose there is a derivation of depth $k > m$. Then at least one symbol, say a,

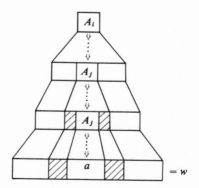

Fig. 3.1

2. INTRODUCTION TO THE THEORY OF CONTEXT-FREE GRAMMARS

in w has depth k. In the production sequence, there must then be a nonterminal symbol A_j which is used at least twice in the production sequence, as shown in Fig. 3.1.

If we now change this production sequence by omitting the shaded parts, we obtain again a legitimate production sequence, yielding a nonempty word w' where the depth of the symbol a in question has been reduced. Continuing in this manner, we can change the production sequence to one of depth at most m. Finally we note that there are only finitely many production sequences of depth at most m, and a check of all of these will reveal effectively whether any of them produces a terminal word.

Our intuitive feeling that nonterminal symbols A_i for which $C_i = \varnothing$ are superfluous in the grammar is verified as follows. Suppose that $C_i = \varnothing$. Let G' arise from G by omitting all production rules in which the symbol A_i occurs.

2.5 Proposition. G and G' generate the same set of words.

Proof. All we need to show is that there is no word that is produced by G but not by G'. If A_i needs to be involved in a production of w, then w will contain nonterminal symbols; otherwise that part of w that arises from A_i would consist of terminal symbols only and $C_i \neq \varnothing$, contrary to assumption.

From Propositions 2.4 and 2.5, it follows that we can make the

Convention. All context-free grammars to be considered have the property that $C_i \neq \varnothing$ for all i.

The question, however, whether $C_i \cap C_j \neq \varnothing$ for $i \neq j$ is a far from trivial task, as is the question of whether the axiom of unique readability holds. Indeed, we see in Section 4 that there is no *general* procedure which would allow us to decide these questions for any given grammar. (But, of course, for some given grammars these questions may turn out to be solvable by this or that method; there is just no *general* method.)

Clearly, every structure G associated to a context-free grammar satisfies the induction axiom. (Why?)

2.6 Definition. A context-free grammar G whose associated structure G is grammatical is called a *nonambiguous* grammar, otherwise G is called *ambiguous*.

2.7 Problems
(a) Show that the grammar $\sigma \to bA$, $\sigma \to aB$, $A \to a$, $B \to b$, $A \to a\sigma$, $B \to b\sigma$, $A \to bAA$, $B \to aBB$ is ambiguous.
(b) The grammar $\sigma \to aAA$, $\sigma \to bA$, $A \to cA$, $A \to c$ is unambiguous.

Let G be a context-free grammar, and let w be a word over the terminal alphabet of G. The following questions arise naturally: Is w a word in the language produced by G? If it is, can we effectively find a production sequence? Can we also find a parsing tree? These questions are answered by uniform effective procedures.

2.8 Proposition. There is a decision procedure for the question of whether $\sigma \stackrel{*}{\Rightarrow} w$, for any word w.

Proof. Observe first that whenever $u \underset{P}{\Rightarrow} v$, the length of v is not smaller than that of u. The only time the length is not actually increased is when a production of the form $A \to b$, b terminal, or $A \to B$, B nonterminal, is employed. For each symbol of w, there are, however, at most $m - 1$ productions of the form $A \to B$ that need to be employed in order to put that symbol into place (where m is the number of nonterminals). From this it is easy to see that $k = m \cdot \text{length}(w)$ is an upper bound on the number of productions needed to derive w. Given w, all we therefore need to do is to check all of the finitely many derivations of length at most k and see whether any one of them produces w. (The proof can be simplified by using Problem 2.9. We return to the problem later.)

2. INTRODUCTION TO THE THEORY OF CONTEXT-FREE GRAMMARS

2.9 Problem. Show that without loss of generality, we may assume that a context-free grammar has no productions of the form $A \to B$, where both A and B are nonterminals.

2.10 Proposition. For any production sequence

$$\sigma = u_0 \Rightarrow u_1 \Rightarrow \cdots \Rightarrow u_n = w$$

of P, we can effectively find a parsing tree for w, and conversely.

The general proof is quite apparent from the treatment of the simple example that follows.

Example. Consider the following context-free grammar over the terminal alphabet $\{a, b, c, +, (,)\}$.

$$P \begin{cases} S \to S+S & \text{syntactical operation } g_1 \\ S \to FF & \text{syntactical operation } g_2 \\ F \to (S+S) & \text{syntactical operation } g_3 \\ F \to FF & \text{syntactical operation } g_4 \\ S \to a, \quad S \to b, \quad S \to c \\ F \to a, \quad F \to b, \quad F \to c \end{cases}$$

There are two syntactical categories, S and F, and four syntactical operations constituting the structure $\langle S, F; g_1, g_2, g_3, g_4 \rangle$. Let $w \in S$ be the word $(a + ab)c$. The following derivation produces w:

$$S \underset{(g_2)}{\Rightarrow} FF \underset{(g_3)}{\Rightarrow} Fc \Rightarrow (S+S)c \underset{(g_2)}{\Rightarrow} (a+S)c$$

$$\Rightarrow (a + FF)c \Rightarrow (a + aF)c \Rightarrow (a+ab)c$$

where we have indicated, whenever applicable, the syntactical operations corresponding to the various productions employed. The parsing tree is found by arranging the symbols for the functions $g_1, \ldots, g_4$ according to their occurrence as annotations to the derivation:

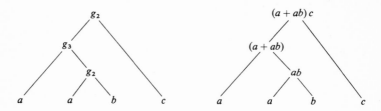

which is obviously a parsing tree of $(a + ab)c$. Conversely, suppose we are given a parsing tree such as this:

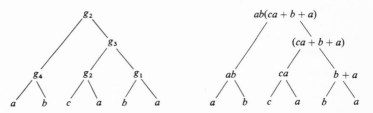

The derivation is found by taking leftmost branches as follows:

$S \Rightarrow FF \Rightarrow FFF \Rightarrow aFF \Rightarrow abF \Rightarrow ab(S+S) \Rightarrow ab(FF+S)$
$\Rightarrow ab(cF+S) \Rightarrow ab(ca+S) \Rightarrow ab(ca+S+S) \Rightarrow ab(ca+b+S)$
$\Rightarrow ab(ca+b+a)$.

This is what is called a leftmost derivation of $ab(ca + b + a)$, since it is always the leftmost nonterminal symbol that is replaced according to a production. If $S \stackrel{*}{\Rightarrow} w$, then w has a leftmost derivation which can be effectively found (see above: first construct the parsing tree according to the given derivation).

2.11 Problem. Show that if a grammar is unambiguous, then every word has at most one leftmost derivation, and vice versa. (This is often taken as a definition for unambiguity for context-free grammar, since it does not appeal to structures and such.)

2. INTRODUCTION TO THE THEORY OF CONTEXT-FREE GRAMMARS

Remark. In the literature it is sometimes customary to label the parsing tree by the syntactical categories rather than by the operations or the expressions generated. For example, the parsing tree would be labeled

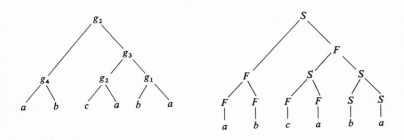

2.12 Definition. A context-free grammar is in (Greibach) *normal form* if all productions are of the form

$$A_i \to a_j w$$

where a_j is a terminal symbol and w is a string (possibly empty) of non-terminal symbols.

2.13 Theorem.[2] For each context-free grammar G, without productions of the form $A_i \to \lambda$, we can effectively find a context-free grammar G^{nf} in normal form such that G and G^{nf} generate the same language.

Proof. The main step in the proof consists in showing

2.14 Lemma. If G is a context-free grammar, then we can effectively find a grammar G'' which generates the same language and whose productions are all of the form $A \to uw$ where $\lambda \neq u \in T^*$ and $w \in N^*$.

---

[2] This theorem is due to Greibach [5].

Let us assume the lemma and prove the theorem. Consider a production

$$A \to u_0 W_1 u_1 W_2 \cdots W_n u_n$$

where $u_0, \ldots, u_{n-1}$ are nonempty elements of T^*, $W_1, \ldots, W_n$ are nonempty elements of N^*, and u_n is a (possibly empty) element of T^*. Now replace this production by two new ones

$$A \to u_0 W_1 C$$
$$C \to u_1 W_2 \cdots W_n u_n$$

where C is a new nonterminal symbol. The new grammar obviously generates the same language as the old one. But in the new production, the number of alternations between strings of terminal symbols and strings of nonterminal symbols has been reduced. This process can of course be repeated as often as necessary for each production in turn, thus resulting in productions which all have the form

$$A \to uW, \quad u \neq \lambda, \quad u \in T^*, \quad W \in N^*.$$

In a next, equally trivial step, we reduce this type of production to productions of the form

$$A \to aW, \quad a \in T, \quad W \in N^*.$$

Namely, take any production

$$A \to a_{i_1} \cdots a_{i_k} W$$

and replace it by the following

$$A \to a_{i_1} B_2$$
$$B_2 \to a_{i_2} B_3$$
$$\vdots$$
$$B_k \to a_{i_k} W$$

where $B_2, \ldots, B_k$ are new nonterminal symbols. This new grammar is in normal form and does generate the same language.

2. INTRODUCTION TO THE THEORY OF CONTEXT-FREE GRAMMARS

Now for the proof of the lemma. We need the following auxiliary concept.

2.15 Definition. Let G be a context-free grammar, $\{A_1,\ldots,A_m\} = N$, $\{a_1,\ldots,a_n\} = T$. The set of nodes of the *graph of G* is $N \cup T$. Two nodes $\textcircled{x}$, $\textcircled{y}$ are connected by an arrow $\textcircled{x} \to \textcircled{y}$ if x is nonterminal and y is the first symbol in a word w such that $x \to w$ is in P.

Example. The graph of

$$S \to S+S, \quad F \to [S+S], \quad S \to q, \quad F \to q$$
$$S \to FF, \quad S \to p, \quad F \to p$$

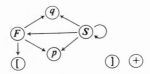

Fig. 3.2

is given in Fig. 3.2. First, we show that we can change the grammar (without changing the language) in such a way that the graph of the new grammar has no cycle $\textcircled{S}\!\curvearrowright$. Namely, suppose that $S \to Su_1, \ldots, S \to Su_r$ is an enumeration of all productions which are responsible for this cycle, and that $S \to v_1, \ldots, S \to v_s$ are the remaining S-productions. We introduce a new nonterminal symbol Z and replace the set of old S-productions by the following set

$$Z \to u_1, \cdots, Z \to u_r, \quad S \to v_1, \cdots, S \to v_s$$
$$Z \to u_1 Z, \cdots, Z \to u_r Z, \quad S \to v_1 Z, \cdots, S \to v_s Z.$$

It is obvious that the new grammar has no cycle at S nor does it have one at the new nonterminal symbol Z. The new grammar clearly generates the same language.

Example (continued).   New S-productions:

$$Z \to +S, \quad S \to FF, \quad S \to p, \quad S \to q$$
$$Z \to +SZ, \quad S \to FFZ, \quad S \to pZ, \quad S \to qZ.$$

Thus, we may assume that the graph of the grammar has no such nodes. The next step is to eliminate cycles involving more than one node. The procedure is in principle the same. Let us consider a cycle of the form shown in Fig. 3.3 where we have indicated the first nodes X, Y, Z, R, S

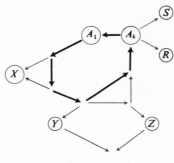

Fig. 3.3

outside all paths that lead back to the link $A_k \to A_1$. We wish to cut at this link. For this purpose we make the following enumerations.

(1) $A_k \Rightarrow A_k u_1, \ldots, A_k \Rightarrow A_k u_r$: all derivations along paths from A_k to A_k using the link only once. (We may assume this set finite since all inner cycles can be considered cut first. In the example above, there are two such paths.)

(2) $A_k \Rightarrow v_1, \ldots, A_k \Rightarrow v_s$: all derivations along paths from A_k to one of $X, T, Z, R, S, \ldots$.

Now let us introduce the new nonterminal symbol B and replace the set of A_k-productions by

$$B \to u_1, \cdots, B \to u_r, \quad A_k \to v_1, \cdots, A_k \to v_s,$$
$$B \to u_1 B, \cdots, B \to u_r B, \quad A_k \to v_1 B, \cdots, A_k \to v_s B.$$

2. INTRODUCTION TO THE THEORY OF CONTEXT-FREE GRAMMARS

In the graph of the new production system, there is no arrow from A_k to A_1. Note also that the introduction of B does not also result in the introduction of new cycles (since there is no production $A_i \to Bw$), nor are there any new arrows introduced apart from those leading out of B.

Thus, after finitely many steps, we obtain a grammar whose graph is cycle free and which generates the same language. Let us now consider a grammar with a cycle-free graph. It consists of components such as that shown in Fig. 3.4. Each of the final nodes must have a terminal symbol as label. (If it had a label A_j, then $C_j = \varnothing$, or $A_j \to \lambda$ which is excluded.) For each link in this graph, there are finitely many productions, and hence for each path from A_i to a final node, say a_j, there are only finitely many words $w_1, \ldots, w_s$ such that $A_i \Rightarrow a_j w_k$ by following this path. Let us replace the set of A_i-productions by the productions $A_i \to a_j w_k$ for all such a_j and w_k. The resulting grammar will generate the same set of words, and it will have a component (Fig. 3.5) in which the maximal length of paths is reduced. Clearly, if we continue this way, we finally obtain a grammar whose graph consists only of the components given in Fig. 3.6, which is what we are after.

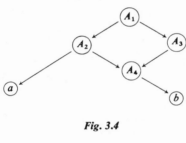

Fig. 3.4

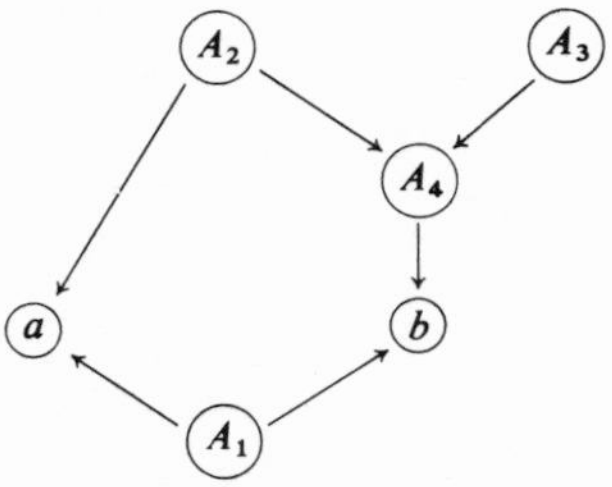

Fig. 3.5

3. ELEMENTARY SYNTAX

Fig. 3.6

Example (continued).   Consider the graph shown in Fig. 3.7.

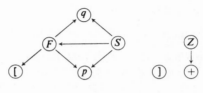

Fig. 3.7

$$\text{path} \quad \underset{\underset{S\to FFZ}{S\to FF}}{\textcircled{S}\longrightarrow}\underset{F\to [S+S]}{\textcircled{F}\longrightarrow}\textcircled{[}\;\bigg\}\;\text{gives}\quad \begin{array}{l} S\to [S+S]F \\ \\ S\to [S+S]FZ \end{array}$$

$$\text{path} \quad \underset{\underset{S\to FFZ}{S\to FF}}{\textcircled{S}\longrightarrow}\underset{F\to q}{\textcircled{F}\longrightarrow}\textcircled{q}\;\bigg\}\;\text{gives}\quad \begin{array}{l} S\to qF \\ \\ S\to qFZ \end{array}$$

$$\text{path} \quad \underset{\underset{S\to FFZ}{S\to FF}}{\textcircled{S}\longrightarrow}\underset{F\to p}{\textcircled{F}\longrightarrow}\textcircled{p}\;\bigg\}\;\text{gives}\quad \begin{array}{l} S\to pF \\ \\ S\to pFZ \end{array}$$

$$\text{path} \quad \underset{\underset{S\to pZ}{S\to p}}{\textcircled{S}\longrightarrow}\textcircled{p}\;\bigg\}\;\text{gives}\quad \begin{array}{l} S\to p \\ \\ S\to pZ \end{array}$$

$$\text{path} \quad \underset{\underset{S\to qZ}{S\to q}}{\textcircled{S}\longrightarrow}\textcircled{q}\;\bigg\}\;\text{gives}\quad \begin{array}{l} S\to q \\ \\ S\to qZ \end{array}$$

2. INTRODUCTION TO THE THEORY OF CONTEXT-FREE GRAMMARS 215

Example. Convert to Greibach normal form:

$$A_1 \to A_2 A_3$$
$$A_2 \to A_3 A_1$$
$$A_3 \to A_1 A_2$$
$$A_2 \to b$$
$$A_3 \to a.$$

The graph is shown in Fig. 3.8. (*Comment*: For convenience we attach labels u, v to the arrows

$$\underset{}{(A_i) \xrightarrow{u,v} (A_j)} \quad \underset{}{(A_i) \xrightarrow{w} (a_j)}$$

to indicate productions $A_i \to A_j u$, $A_i \to A_j v$, $A_i \to a_j w$, for example.)

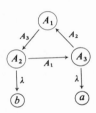

Fig. 3.8

Cut link $(A_3) \longrightarrow (A_1)$:

$$A_3 \Rightarrow A_3 \underbrace{A_1 A_3 A_2}_{u}; \quad A_3 \Rightarrow \underbrace{b A_3 A_2}_{v_1}, \quad A_3 \Rightarrow \underbrace{a.}_{v_2}$$

Result:

$$A_1 \to A_2 A_3$$
$$A_2 \to A_3 A_1$$
$$Z \to A_1 A_3 A_2$$
$$Z \to A_1 A_3 A_2 Z$$

$$A_3 \to bA_3A_2$$
$$A_3 \to a$$
$$A_3 \to bA_3A_2Z$$
$$A_3 \to aZ$$
$$A_2 \to b.$$

The resulting graph is depicted in Fig. 3.9.

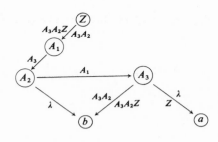

Fig. 3.9

Productions:

$$A_3 \to a \qquad Z \to aA_1A_3A_3A_2$$
$$A_3 \to aZ \qquad Z \to aZA_1A_3A_3A_2$$
$$A_3 \to bA_3A_2 \qquad Z \to bA_3A_2A_1A_3A_3A_2$$
$$A_3 \to bA_3A_2Z \qquad Z \to bA_3A_2ZA_1A_3A_3A_2$$
$$A_2 \to aA_1 \qquad Z \to bA_3A_3A_2$$
$$A_2 \to aZA_1 \qquad Z \to aA_1A_3A_3A_2Z$$
$$A_2 \to bA_3A_2A_1 \qquad Z \to aZA_1A_3A_3A_2Z$$
$$A_2 \to bA_3A_2ZA_1 \qquad Z \to bA_3A_2A_1A_3A_3A_2Z$$
$$A_2 \to b \qquad Z \to bA_3A_2ZA_1A_3A_3A_2Z$$
$$A_1 \to aA_1A_3 \qquad Z \to bA_3A_3A_2Z.$$
$$A_1 \to aZA_1A_3$$

$$A_1 \to bA_3A_2A_1A_3$$
$$A_1 \to bA_3A_2ZA_1A_3$$
$$A_1 \to bA_3$$

2.16 Problem. A context-free grammar is said to be in Chomsky normal form if all productions are of the form

$$A_i \to A_j A_k$$

or

$$A_i \to a_p.$$

Show how to obtain a grammar in Chomsky normal form which generates the same language as a given context-free grammar.

3. MEMORY MANAGEMENT

We now return to the basic problem touched upon in the introduction to this chapter. Namely, how can a machine read and "understand" a formal program? A full treatment of this question would involve too much for us here, so we shall be content with one small, but important, aspect.

Imagine that we have written a program and feed it now, symbol after symbol, into a computer. Will the computer *accept* it as well formed? Indeed, what is involved as regards operations and decisions which the computer is to be capable of for it to be up to this task?

As a first, simplifying, assumption, let us postulate that the set of all well-formed programs is a context-free language. (See Section 1.) Let T be the terminal alphabet, N the nonterminal alphabet of the corresponding context-free grammar. The input would therefore consist of a word over T. The computer would have a number of memory cells in which are stored words over T or some extension of that alphabet. The acceptance program then would perform some operations on the input and the memory contents. Of course, there are very varied kinds of such programs imaginable.

We discuss only one system of memory management here, namely the *push-down-store* or *stack*.[3] All the stored information is located at only one address y, and access to this information is very limited. The contents of y form a word over a special alphabet, the stack alphabet, and we have only the following capabilities with respect to y:

(i) $y := S$, where S is a particular element of the stack alphabet.
(ii) $y := \gamma_i(y)$, $i = 1, \ldots, k$, where for each γ_i we are given a word w_i over the stack alphabet and $\gamma_i(w)$ arises from w by replacing the leftmost symbol of w by the word w_i; if w is empty, then so is $\gamma_i(w)$.
(iii) test(y). This is a routine which has one or more exits for each symbol in the stack alphabet, plus one exit for emptiness; it tests the leftmost symbol of the contents of y.

The environment in which the push-down-store works varies; for the present purposes, we are content with an *input location* x which contains words over the input alphabet and with respect to which we assume the following capabilities:

(iv) test(x)
(v) $x := d(x)$

where $d(w)$ results from deleting the leftmost symbol of w, if any; $d(\lambda) = \lambda$.

3.1 Definition. A weak *push-down-store acceptor* is a program over the capabilities (i)–(v) above, which has the form shown in Fig. 3.10.

A weak push-down-store acceptor is *nondeterministic* if $\boxed{\text{test}(x)}$ or $\boxed{\text{test}(y)}$ has more than one exit for some symbol in their respective alphabets, otherwise the acceptor is called deterministic.

Convention. In the present section we drop the adjective "weak" from "weak push-down-store acceptor."

---

[3] The connection between push-down-stores and languages was first explored by Samelson and Bauer [6].

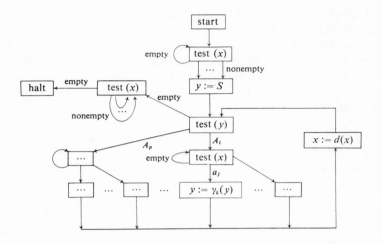

Fig. 3.10

The set of words over the input alphabet *accepted* by the push-down-store acceptor consists of those words on which the program terminates. We show that such sets are exactly the context-free languages.

TABLE 3.1

Push-down-store	Input
S	$p + [q + p][p + p]$
$+S$	$+ [q + p][p + p]$
S	$[q + p][p + p]$
FF	$q + p][p + p]$
$+S]F$	$+ p][p + p]$
$S]F$	$p][p + p]$
$]F$	$][p + p]$
F	$[p + p]$
$S + S]$	$p + p]$
$+S]$	$+ p]$
$S]$	$p]$
$]$	$]$
Empty	Empty

3. ELEMENTARY SYNTAX

Example. Consider the language produced by the grammar of our example in Section 2. We show in Table 3.1 how a word $p + [q + p][p + p]$ of that language is being accepted by a push-down-store acceptor. Roughly, the content of the push-down-store indicates the "unfinished business" in processing the input sequence. Note also the correlation between productions of the grammar and subsequent words in the push-down-store.

3.2 *Theorem.* A set of words over a finite alphabet is accepted by a nondeterministic push-down-store acceptor if and only if it is a context-free language.

Proof. Suppose that we are given a context-free language. Its grammar may be assumed, without loss of generality, to be in normal form. We construct a push-down-store acceptor whose stack alphabet N consists of the nonterminal symbols of the grammar (and whose input alphabet T, of course, consists of the terminal symbols of the grammar). For each production

$$A_l \to a_j W_k \qquad (a_j \in T, \quad W_k \in N^*)$$

of the grammar we stipulate a capability γ_k such that $\gamma_k(w) = W_k w$ for all $w \in N^*$. As the distinguished element of the stack alphabet, we choose the start symbol S of the grammar. The program is obtained by providing a link, as shown in Fig. 3.11, for each such production. It is clear that this acceptor halts on an input word exactly if it is produced by the grammar.

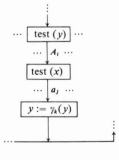

Fig. 3.11

3. MEMORY MANAGEMENT

Conversely, if we are given a push-down-store acceptor, the grammar (in normal form) is simply read off the links of the programs.

Remark. We may slightly generalize the notion of a push-down-store acceptor by introducing the additional indeterminacy, shown in Fig. 3.12. That is, we offer the alternative, each time that we go around the loop, to decrement the contents of x or leave it constant. (This still falls short of what is called a nondeterministic push-down-store acceptor in the literature.)

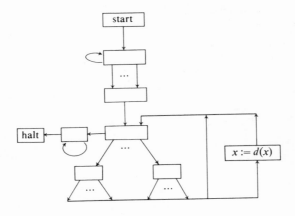

Fig. 3.12

3.3 Problems

(a)* Show that the generalization shown in Fig. 3.12 does not extend the class of sets acceptable by such devices.

(b) Construct a push-down-store acceptor for the language generated by the following grammar:

$$A \to BB$$
$$A \to aA$$
$$B \to bAB$$
$$B \to c$$

(start symbol A)

3. ELEMENTARY SYNTAX

3.4 Example. Find a push-down-store acceptor for the following language:

$$L = \{a^n b^n : n = 1, 2, \ldots\}.$$

Solution. The following is a grammar for L:

$$S \to aSb$$
$$S \to ab$$

Greibach normal form:

$$S \to aSB$$
$$B \to b$$
$$S \to aB$$

Acceptor: See Fig. 3.13.

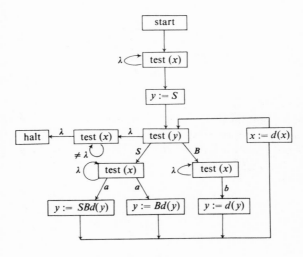

Fig. 3.13

The history of *aaabbb* is shown in Table 3.2 where x and y form the contents when entering the loop.

TABLE 3.2

x	y
aaabbb	S
aabbb	SB
abbb	SBB
bbb	BBB
bb	BB
b	B
λ	λ

4.* UNSOLVABILITY OF GRAMMATICAL PROBLEMS

Let there be given a finite alphabet $A = \{a, b\}$, say, and a finite list of pairs of words

$$\langle x_1, y_1 \rangle, \langle x_2, y_2 \rangle, \ldots, \langle x_n, y_n \rangle$$

over A. Let us consider the following question.

Post Correspondence Problem. Does there exist a finite sequence $i_1, \ldots, i_m$ of indices $i_k \in \{1, \ldots, n\}$ such that the words

$$x_{i_1} x_{i_2} \cdots x_{i_m} \quad \text{and} \quad y_{i_1} y_{i_2} \cdots y_{i_m}$$

are identical?

Examples
(a) The problem $\langle ba, a \rangle$, $\langle b, bb \rangle$ has solution 2, 1: bba is the same as bba.

(b) The problem $\langle ba, ab \rangle$, $\langle a, b \rangle$ has no solution: If it had, then x_{i_1} should be an initial part of y_{i_1} or y_{i_1} an initial part of x_{i_1}. But neither is the case for any of the two pairs.

4.1 Theorem.[4] There is no recursive decision procedure for Post's correspondence problem for alphabets with at least two symbols.

(The proof of this theorem is not given in these notes; it proceeds by reducing the correspondence problem to a halting problem.)

4.2 Theorem.[5] There is no decision procedure for the ambiguity problem for context-free grammars.

Proof. The proof is accomplished by reduction to the correspondence problem. That is, suppose we have a solution to the ambiguity decision problem and we are given a correspondence problem over $A = \{a, b\}$ by $\langle x_1, y_1 \rangle, \ldots, \langle x_n, y_n \rangle$. We shall construct a context-free grammar R such that R is ambiguous exactly when this correspondence problem has a solution. (Thus, if we had a decision procedure for ambiguity, we would then also have one for the correspondence problem!)

Let us encode the letters a, b into the alphabet $\{0, 1\}$ by writing "1" for "a" and "11" for "b". This encoding is extended to $\{a, b\}^*$ by separating the 1's and 11's by 0's. The word *abba*, for example, is encoded as 101101101. We denote the encoded word w by $\bar{w}$. Thus the formal definition would read

$$\bar{\lambda} = \lambda$$
$$\overline{wa} = \bar{w}01, \qquad \overline{wb} = \bar{w}011.$$

In a similar way, we encode natural numbers k by sequences of $k + 1$ numerals "1," thus

$$\bar{k} = \underbrace{11 \cdots 1}_{k+1}.$$

---

[4] Post showed the unsolvability of the correspondence problem in 1946; for a textbook presentation of a proof due to Floyd see Yasuhara [7, pp. 49–51].

[5] This theorem was found independently by Cantor, Floyd, Chomsky, and Schützenberger; see, for example, Floyd [4].

4. UNSOLVABILITY OF GRAMMATICAL PROBLEMS

Consider now the following grammars:

$$R_x \begin{cases} \sigma \to 0\bar{k}00\bar{x}_k, & k = 1, 2, \ldots, n; \\ \sigma \to 0\bar{k}\sigma\bar{x}_k, & k = 1, 2, \ldots, n. \end{cases}$$

$$R_y \begin{cases} \tau \to 0\bar{k}00\bar{y}, & k = 1, 2, \ldots, n; \\ \tau \to 0\bar{k}\tau\bar{y}_k, & k = 1, 2, \ldots, n. \end{cases}$$

Both R_x and R_y are clearly unambiguous. Let us combine R_x and R_y to obtain the following grammar

$$R \begin{cases} \rho \to \sigma \\ \rho \to \tau \\ R_x, \quad R_y \end{cases}$$

R is potentially ambiguous; it is ambiguous iff a word $w \in \{0,1\}^*$ is derivable in both R_x and R_y. Thus R is ambiguous iff

$$R_x \cap R_y \neq \varnothing.$$

Now $R_x \cap R_y \neq \varnothing$ iff there is a sequence $k_1, \ldots, k_m$ such that

$$0\bar{k}_1 0 \cdots 0\bar{k}_m 00\bar{x}_{k_m}\bar{x}_{k_{m-1}} \cdots \bar{x}_{k_1} = 0\bar{k}_1 0 \cdots 0\bar{k}_m 00\bar{y}_{k_m}\bar{y}_{k_{m-1}} \cdots \bar{y}_{k_1},$$

which is the case iff there is a sequence $k_1, \ldots, k_m$ for which

$$\bar{x}_{k_1} \cdots \bar{x}_{k_m} = \bar{y}_{k_1} \cdots \bar{y}_{k_m};$$

that is, iff

$$x_{k_1} \cdots x_{k_m} = y_{k_1} \cdots y_{k_m}$$

for some sequence $k_1, \ldots, k_m$. It follows that R is ambiguous iff the given correspondence problem has a solution.

4.3 Problem. There is a decision procedure for correspondence problems over the alphabet $\{1\}$. Find it.

REFERENCES

Books

[1] S. Ginsburg, *The Mathematical Theory of Context-Free Languages.* McGraw-Hill, New York, 1966.
[2] J. E. Hopcroft and J. D. Ullman, *Formal Languages and Their Relation to Automata.* Addison-Wesley, Reading, Massachusetts, 1969.

Original Articles

[3] N. Chomsky, Formal properties of grammars. *In Handbook of Mathematical Psychology*, Vol. 2 (R. D. Luce, R. R. Bush, and E. Galanter, eds.). Wiley, New York, 1963, pp. 323–418.
[4] R. W. Floyd, On ambiguity in phrase structure languages. *Comm. ACM*, **5**, 526, 534 (1962).
[5] S. A. Greibach, A new normal form theorem for context-free phrase structure grammars. *J. Assoc. Comput. Mach.*, **12**, 42–52 (1965).
[6] K. Samelson and F. L. Bauer, Sequential formula translation. *Comm. ACM*, **3**, 76–82 (1960).
[7] A. Yasuhara, *Recursive Function Theory and Logic.* Academic Press, New York, 1971.

Index

A

Accepted word, *see* Finite automata, Push-down-store acceptor
Accessible state, 61
Accumulator, 173
Ambiguous, *see* Grammar
Arbib, M. A., 84
Atom, 197
Automata, *see* Finite automata
Axiom, 21, *see also* Free monoid, Grammatical structure, Induction, Unique readability

B

Bauer, F. L., 218, 226
Biconditional (truth function), 76
Boole, G., 76, 85
Bounded function, 143
Bounded halting problem, 170
Büchi, J. R., 30, 85

C

Category, *see* Syntactical category
Chomsky, N., 25, 36, 85, 202, 226
Church's thesis, 164
Complete, *see* Functionally complete
Composition
 of functions, 100
 of programs, 115
Computation, *see* RASP, Turing machine, Universal calculator
Computer, *see* Finite transducer

Computing time (RASP), 186
Concatenation, 5, 9
Conditional (truth function), 75
 instruction, 92
Congruence relation, 60
 class, 60
Conjunction (truth function), 75
Connected part (of finite transducer), 61
Cycle (of graph), 211

D

Davis, M., 164, 170, 189
Decidable, *see* Recursively decidable
Decision problem
 ambiguity problem for context-free grammars, 224
 halting problem, 161
 Post correspondence problem, 223
Decision procedure, 163
Decision program, 161
Deduction, *see* Production sequence
Delay line, 72
Derivation, *see* Production sequence
Diagonal function, 149
Disjoint union (of partial functions), 114, 131
Disjunction (truth function), 75
Disjunctive normal form, 84
Distance function, 108

E

Elgot, C. C., 173, 190
Entrance (to program), 116

Equality function, 108
Equivalence class, 62
Equivalence relation, 60
Equivalent, *see* Finite automata, Finite transducer, Grammar, Program, State
Exit (halt instruction), 114

F

Finite automata, 30–55
  accepted set, 33, 46
  accepted word, 33, 46
  alphabet, 30, 46
  deterministic, 47
  equivalent, 69
  final state, 30
  nondeterministic, 46–53
  start state, 30
  transition function, 30
  transition relation, 46
Finite computer, *see* Finite transducer
Finite transducer, 57–69
  computable function, 58
  equivalent, 59
  input alphabet, 57
  minimal, 62
  output alphabet, 57
  output assignment, 57
  transition function, 57
Floyd, R. W., 224, 226
Free monoid, 11–15
  axioms, 11
Function, *see also* names of specific functions
  bounded, 143
  characteristic (of exit), 128
  composition of, 100
  computable, *see* Finite transducer, RASP, Turing machine, Universal calculator
  domain of, 6
  general recursive, 171
  majorized, 143

next-action (Turing machine), 166
next-state (Turing machine), 166
partial, 97
partial recursive, 110, 114
primitive recursive, 106, 110, 137–149
projection, 77, 106
range of, 6
recursive, *see* General recursive
total, 97
transition, *see* Finite automata, Finite transducer
Functionally complete, 82

G

Generate, free monoid, 15
  set of truth functions, 77
General recursive function, *see* Function
Generation sequence, 78
Generator,
  of monoid, 11, 15
  of truth functions, 78
Ginsburg, S., 226
Ginzburg, A., 84
Gödel numbers, 150–152, 167, 168
Grammar, 16, 200
  ambiguous, 206, 208, 224
  associated structure, 203
  context-free, 199–225
  context-sensitive, 202
  equivalent, 27
  left-linear, 26
  nonterminal alphabet, 25, 200
  production, 25, 200
  rewrite rule, 16, *see also* Production
  right-linear, 25, 203
  start symbol, 25
  terminal alphabet, 25, 200
  type of, 202
Grammatical structure, 197
Graph
  of function, 13
  of grammar, 28, 211
Greibach, S. A., 209, 226

H

Halting problem, 161
Harrison, M. A., 84
Hartmanis, J., 84, 173, 184
Homomorphism
  between finite transducers, 61
  between monoids, 13
Hopcroft, J., 85

I

Induction
  axiom for free monoids, 11
  for grammatical structures, 197
  on length of words, 6
Initial functions, 106
Input, see Finite transducer, RASP, Turing machine, Universal calculator
  channel, 70
  variable, 97
Instruction, see RASP, Universal calculator
Isomorphism
  between finite transducers, 63
  between monoids, 14
Iteration, 132, 144

J

Jam, 173

K

Kleene, S. C., 38, 85, 90, 110, 170, 189, 190
Kleene's normal form theorem, 159, 160
Korfhage, R. R., 76, 85

L

Label, 92, 196
Language
  generated by a grammar, 200
  left-linear, 26
  regular, 29
  right-linear, 25

Loop complexity, 139
Loop program, 138–149
Looping of program, 116

M

McCarthy, J., 84
McCulloch, W. S., 31, 85
Majorized function, 143
Markov, A. A., 90, 189
Memory registers, 173
Meyer, A. R., 138, 190
Miller, G. R., 25, 36, 85
Minimalization
  scheme of, 109
  of finite transducer, 63–69
Minsky, M., 84, 189
Modified difference function, 107
Monoid, 10
  free, 11–15
Moore, E. F., 58, 84
μ-scheme, 109

N

Negation (truth-function), 75
Nelson, R. J., 84
Nerode, A., 58, 85
Next-state function, see Transition function
Normal form
  Chomsky, 217
  disjunctive, 84
  Greibach, 209–217
  for programs, 118–127
  theorem, see Kleene's normal form theorem

O

Output, see Finite transducer, RASP, Turing machine, Universal calculator
  channel, 70
  variable, 97
Operator, 90

P

Partial recursive function, *see* Function
Partition, 60
Peter, R., 190
Pitts, W., 31, 85
Post correspondence problem, 223–224
Post, E. L., 74, 85, 224
Predecessor function, 106
Primitive recursive function, *see* Function
Product function, 107
Production, 25, 200
Production sequence, 35, 200
Program
  decision, 161
  equivalent, 121
  fixed, 178
  for RASP, 176
  for universal calculator, 93, 196
  self-modifying, 178
  stored, 149–160
  universal, 154, 162
Projection function, 77, 106
Power function, 107
Push-down-store acceptor, 218–223
  accepted set, 219
  deterministic, 218
  weak, 218

R

Rabin, M. O., 46, 85
Random-access stored-program machine, *see* RASP
RASP, 172–189
  computable, 179–180
  computation, 173
  input, 176
  instruction, 174–176
  output, 176
  program, 176
Recursion, equation, 8, 102
  primitive, 7, 104
  simultaneous, 105, 110–113
  scheme of, 104

Recursive, *see* Recursively decidable function, *see* Function
  set, 160
Recursively decidable, 160
  enumerable, 172
Regular, expression, 43
  language, 29
  set, 38
Ritchie, D. M., 138, 190
Ritchie, R. W., 41, 85
Robinson, A., 173, 190
Rogers, H., 190

S

Salomaa, A., 44, 85
Samelson, K., 218, 226
Scanned symbol (Turing machine), 166
Scott, D., 46, 85
Semantics, 193
Semigroup, 10
Set, accepted, *see* Finite automata, Push-down-store acceptor
  generated by grammar, 200
  recursive, 160
  recursively enumerable, 172
  regular, 38
Shannon, C. E., 84
Sheffer's function (truth function), 76
Sheperdson, J. C., 89, 190
Signum function, 108
State, *see also* Finite automata, Finite transducer, Turing machine
  accessible, 61
  equivalent, 59
Sturgis, H. E., 89, 190
Syntactical category, 195–196
  operation, 195–196
Syntax, 193
Symbol, 3, 4, 11

T

Tape, *see* Turing machine
Transition, diagram, 31
  function, 30, 57
  relation, 46

Truth function, 74–84, *see also* specific truth functions
value, 74
Turing, A., 164, 190
Turing machine, 165–170
  computable function, 169
  computation, 166
  configuration, 166
  input, 166
  output, 167
  state, 165
  tape, 166
Turing's thesis, 170

U

Ullman, J. D., 226
Undecidable, 160
Unique readability axiom, 197
Universal calculator, 90–137
  computable function, 98
  computation, 94
  input, 94
  instruction, branching, *see* Conditional
  conditional, 92, 196
  halt, 93, 196
  operational, 92, 196
  start, 93, 196
  output, 94
  output variable, 97
  program, 93, 196
  termination, 94
Universal program, 154–159, 162–163

V

Variable, 92

W

Wang, H., 89, 190
Word, accepted, *see* Finite automata, Push-down-store acceptor
  empty, 5
  length of, 6

Y

Yasuhara, A., 190, 224, 226

Z

Zero function, 106

Computer Science and Applied Mathematics
A SERIES OF MONOGRAPHS AND TEXTBOOKS

Editor
Werner Rheinboldt
University of Maryland

Hans P. Künzi, H. G. Tzschach, and C. A. Zehnder
NUMERICAL METHODS OF MATHEMATICAL OPTIMIZATION: WITH ALGOL AND FORTRAN PROGRAMS, CORRECTED AND AUGMENTED EDITION, 1971

Azriel Rosenfeld
PICTURE PROCESSING BY COMPUTER, 1969

James Ortega and Werner Rheinboldt
ITERATIVE SOLUTION OF NONLINEAR EQUATIONS IN SEVERAL VARIABLES, 1970

A. T. Berztiss
DATA STRUCTURES: THEORY AND PRACTICE, 1971

Azaria Paz
INTRODUCTION TO PROBABILISTIC AUTOMATA, 1971

David Young
ITERATIVE SOLUTION OF LARGE LINEAR SYSTEMS, 1971

Ann Yasuhara
RECURSIVE FUNCTION THEORY AND LOGIC, 1971

James M. Ortega
NUMERICAL ANALYSIS: A SECOND COURSE, 1972

G. W. Stewart
INTRODUCTION TO MATRIX COMPUTATIONS, 1973

Chin-Liang Chang and Richard Char-Tung Lee
SYMBOLIC LOGIC AND MECHANICAL THEOREM PROVING, 1973

C. C. Gotlieb and A. Borodin
SOCIAL ISSUES IN COMPUTING, 1973

Erwin Engeler
INTRODUCTION TO THE THEORY OF COMPUTATION, 1973

In preparation

F. W. J. Olver
ASYMPTOTICS AND SPECIAL FUNCTIONS

QA267.5.S4 E53 1973
Engeler / Introduction to the theory of computatio